Science and Policy: Interbasin Water Transfer of Aquatic Biota

Compiled and edited by Jay A. Leitch and Mariah J. Tenamoc

Institute for Regional Studies,
North Dakota State University, Fargo.

Institute for Regional Studies, North Dakota State University
Fargo, North Dakota 58105

International Standard Book Number (ISBN) 0-911042-54-7.

Library of Congress Control Number 2001088265

Table of Contents

List of Figures

Acknowledgements

The Interbasin Water Transfer Study Program (IWTSP) was guided by the Technical Advisory Team (TAT), whose members have served in voluntary capacity. We are grateful and appreciative of their dedication and help in keeping this program focused on science. Members include the following:

Dr. Jay A. Leitch, Chair, Agricultural/Resource Economist, NDSU (May 1988 - Dec 1997)

Dr. Bruce Barton, Biologist, UND (May 1990 - Dec 1990)

Dr. Frank Beaver, Civil Engineer, UND (May 1988 - May 1990)

Dr. Charles Bigelow, Chemist, Univ. of Manitoba (May 1988 - Oct 1990)

Dr. Mario Biondini, Systems Ecologist, NDSU (May 1988 - Dec 1997)

Dr. Gary Clambey, Ecologist, NDSU (May 1988 - Dec 1997)

Dr. Brenda Hann, Zoologist, Univ. of Manitoba (Mar 1992 - Dec 1997)

Mr. Eugene Krenz, Ex Officio, ND State Water Commission (May 1988 - Feb 1996)

Dr. Greg McCarthy, Ex Officio, Chemist, NDSU (Sep 1991 - Dec 1997)

Dr. Rick Nelson, Ex Officio, Biologist, Bureau of Reclamation (Oct 1989 - Dec 1997)

Dr. G. Padmanabhan, Civil Engineer, NDSU (May 1988 - Dec 1997)

Dr. Shahin Rezania, Civil Engineer, UND (May 1990 - Aug 1991)

Dr. Rodney Sayler, Biologist, UND (May 1988 - May 1990)

Dr. Kenneth Stewart, Zoologist, Univ. of Manitoba (Oct 1990 - Dec 1997)

Dr. Charles Turner, Civil Engineer, UND (May 1988 - May 1990)

Dr. Gerald Van Amburg, Biologist, Concordia College (May 1988 - Dec 1997)

We also thank Dr. Ross Collins and Carol Stark, for their patient assistance in getting this volume into final form.

Sponsors

Financial support for the Interbasin Water Transfer Study Program was provided by

Garrison Diversion Conservancy District
Box 140
Carrington, North Dakota U.S.A. 58421

North Dakota State Water Commission
900 East Boulevard
Bismarck, North Dakota U.S.A. 58505

U.S. Bureau of Reclamation
P.O. Box 1017
Bismarck, ND U.S.A. 58502

Additional matching support was provided by parent universities and home institutions of the principal investigators through personnel and equipment cost sharing and waivers of indirect cost recovery. Cooperating institutions were

North Dakota State University, Fargo
University of North Dakota, Grand Forks
University of Manitoba, Winnipeg
Department of Fisheries and Oceans, The Freshwater Institute, Winnipeg

We wish to thank those who contributed to the study. An effort of this scope must rely on the help and contributions of many individuals and institutions. We hope we have reached your expectations. Any errors of commission or omission belong with the authors.

Garrison Diversion Conservancy District[1]

Board of Directors

Members in 1997	City	County
Adams, Bruce (1995-1998)	Minot	Ward
Anderson, Rick (1987-1998)	Coleharbor	McLean
Ashley, Stephen (1991-1998)	Voltaire	McHenry
Christianson, Paul (1983-1998)	Glenburn	Renville
DeKrey, Lester R. (1985-2000)	Valley City	Barnes
Dushinske, Russell L. (1979-1998)	Devils Lake	Ramsey
Fugleberg, Richard H. (1988-2000)	Portland	Steele
Haak, Norman (1987-2000)	Oakes	Dickey
Johnson, David B. (1987-1998)	Minnewaukan	Benson
Johnson, LeRoy (1983-1998)	Rugby	Pierce
Johnson, Lester (1997-1998)	Walcott	Richland
Johnson, Roger (1977-2000)	Tolna	Nelson
Leininger, Kenneth (1981-2000)	Binford	Griggs
Lindgren, Jon (1991-1998)	Fargo	Cass
Lochow, Milton (1987-1998)	Jamestown	Stutsman
Lyons, Warren (1997-2000)	Lisbon	Ransom
Metzger, Steve (1983-2000)	Carrington	Foster
Orn, Maurice (1991-2000)	Stirum	Sargent
Orthmeyer, Frank B. (1989-2000)	Grand Forks	Grand Forks
Reiswig, Tilmer (1973-2000)	McClusky	Sheridan
Richter, Charles (1969-2000)	New Rockford	Eddy
Roble, Jerrold (1997-1998)	Harvey	Wells
Rogers, Kenny (1997-2000)	Maxbass	Bottineau
Sprynczynatyk, Connie (1988-2000)	Bismarck	Burleigh
Strand, Robert (1979-1998)	Portland	Traill
Wendel, Dennis (1997-2000)	LaMoure	LaMoure
Past Members		
Anderson, Lester (1961-1996)	Minot	Bottineau
Berstler, Royal (1979-1982	Jamestown	Stutsman
Dean, John S. (1961-1988)	Hatton	Grand Forks
Eaton, J.C., Jr. (1973-1994)	Minot	Ward
Froemke, Argil R. (1969-1996)	Lisbon	Ransom

[1]The GDCD was established in 1956. GDCD board members represent the 26 counties in the district. Prior to 1988, there were 25 counties represented by the board. See <www.garrisondiv.org>.

Garrison Diversion Conservancy District[1] (cont)

Gottschalk, Forest (1955-1973)	Oakes	Dickey
Harmon, Ralph (1955-1984)	Carrington	Foster
Holand, Roy (1955-1985)	LaMoure	LaMoure
Jordheim, Selmer N. (1971-1990)	Walcott	Richland
Meidinger, R.E. (1973-1978)	Jamestown	Stutsman
Miller, William (1955-57; 1970-80)	Valley City	Barnes
Mork, Kendall (1967-1978)	Hatton	Traill
Nathan, Herbert (1983-1986)	Turtle Lake	McLean
Palmer, Earl (1971-1982)	Glenburn	Renville
Rehovsky, Louis (1973-1984)	Oakes	Dickey
Rygg, Harold E. (1985-1988)	Portland	Steele
Sherva, M. Blake (1967-1970)	Wahpeton	Richland
Shockman, Thomas C. (1985-1996)	LaMoure	LaMoure
Simmers, Francis (1957-1973)	Jamestown	Stutsman
Slotten, Russell (1969-1973)	Fargo	Cass
Sturlaugson, Vernon (1960-1979)	Larimore	Benson

Foreword

Written by:

Former Governor William L. Guy, North Dakota, U.S.A.
Former Garrison Focus Office Coordinator Robert N. Clarkson, Manitoba, CANADA

Foreword
William L. Guy

"Hold back your upstream flood water!" was, for decades, the agonized cry from downstream on the mighty Missouri River as the upstream spring snow melt came surging down across cropland and bank-hugging towns.

Ideas for controlling the Missouri ranged from huge dams across the mainstream, championed by the U.S. Army Corps of Engineers, to consuming water in giant irrigation projects upstream, advocated by the U.S. Bureau of Reclamation. The ideas called for sacrifice and loss to some if others were to gain. The political pressures for controlling the Missouri River covered such a vast constituency in parts, or all, of ten states drained by the giant river that for years the solutions to downstream flood problems were at an impasse.

Finally, in 1944, President Franklin Roosevelt ordered the Army Corps of Engineers and the Bureau of Reclamation to jointly create a Missouri River basin-wide plan that ultimately featured scores of projects, large and small, to provide for beneficial use of water, mitigate flood damage, and recoup economic sacrifice. The Pick/Sloan Missouri Basin Program was born in the Flood Control Act of 1944.

The Flood Control Act was all encompassing in stating the kinds of beneficial uses of water to be developed. It was even specific in calling for 1,008,000 acres of irrigation in North Dakota. Promises of beneficial uses of water upstream were necessary to dampen the opposition to permanently flooding 1.8 million acres of choice bottom land in Wyoming, Montana, North Dakota, and South Dakota so that 1.2 million acres of flood prone land in Iowa, Nebraska, Kansas, and Missouri could become flood free.

Due to the limited sources of water in the western states in the Missouri River basin, the congressional representatives were very sensitive to what the Pick/Sloan Program might do for them—or to them. U.S. Senator Joseph O'Mahoney, a Wyoming Democrat, and U.S. Senator Eugene Millikin, a Colorado Republican, foresaw the possibility of the downstream navigation being the tail that would wag the hundreds of other beneficial use projects that made up the dog. They secured an amendment to the Flood Control Act of 1944 specifically relegating navigation to the bottom of the list of water use priorities. The Army Corps of Engineers, with downstream political pressure and upstream acquiescence, ignored the O'Mahoney/Millikin amendment from the beginning.

Mainstream dam projects were built first, with solid political backing for funding in Congress from all ten Pick/Sloan states. A Missouri Basin Interagency Committee was formed from federal agencies, dominated by the Corps of Engineers and the Bureau of Reclamation, but with state representation invited to sit at the table and listen. A basin-wide Missouri State's Committee of Governors was formed to focus pressure on Congress and the Administration to move authorizations and appropriations along. In the beginning, it was an active and effective committee of governors, with downstream and upstream states working closely together.

By the mid 1960s, six big dams—Fort Peck in Montana; Garrison in North Dakota; Oahe, Fort Randal, and Big Bend in South Dakota; and Gavin's Point on the South Dakota-Nebraska border—were in place, and their reservoirs were soaking up upstream flood water. Finally, downstream states were relieved of the spring devastation from upstream snow melts and spring rains. Then, almost overnight, the Missouri River States' Committee of Governors began to fall apart.

The downstream states had their flood control. The Corps of Engineers had stabilized the river banks and dredged a nine-foot navigation channel from St. Louis, Missouri, to Sioux City, Iowa. The downstream states could export by river navigation. The Missouri River was being managed with Gavin's Point releases so that a constant level of water guaranteed not only navigation and mid-continent grain export but city and industrial water as well. An in-stream flow was assured in the Missouri River that allowed Nebraska to use nearly all the Platte River water for irrigation with only a trickle returning to the Missouri. The upstream flood water, safely stored behind the six big dams, spun hydro power as it was released through turbines. Huge blocks of cheap hydro electricity moved across the Bureau of Reclamation transmission lines following the other benefits to downstream consumers. The agonized cry from downstream of "Hold back your upstream flood waters!" changed to "Hold back our navigation water 'till we need it!'" Ownership of flood water had suddenly been reversed.

No longer were the downstream states interested in completing the ten-state Pick/Sloan Program. The Missouri River States' Committee of Governors dissolved when the self-interest of downstream governors turned from lethargy to hostility. Clearly, downstream states had their benefits and would go to any length to prevent upstream states from consuming Missouri River water.

Downstream hostility showed then and still shows in many ways. In 1979, Nebraska, in concert with the National Wildlife Federation, brought an Endangered Species Act lawsuit before a friendly downstream federal judge to prevent the Missouri Basin Power Project from building a large coal-fired electric generating plant at Wheatland, Wyoming. The plant would annually consume 25,000 acre feet of Wyoming's Laramie River water which would otherwise flow to the Platte River and on to Nebraska irrigators. The downstreamers won their lawsuit and forced the MBPP to limit water consumption and establish a $7.5 million Platte River Whooping Crane Critical Habitat Maintenance Trust Fund to foster habitat for non-existent whooping cranes. Nebraska irrigators and the flyway waterfowl hunting industry hid behind the Endangered Species Act in that lawsuit.

A few years later, Iowa, Missouri, and Nebraska again went before a friendly downstream federal judge and were able to block South Dakota from diverting and selling 50,000 acre feet of its Missouri River water annually, via a pipeline, to the Wyoming coal fields for sluicing coal to Louisiana. Railroads joined the downstream states in preventing consumption of upstream water.

While downstream states were reaping huge benefits from flood control in the form of river navigation, cheap hydro power, and central flyway duck hunting, North Dakotans looked at 840 square miles of their best bottom lands permanently flooded and wondered if Pick/Sloan program benefits would arrive in significant amounts to cut their losses.

North Dakota received its first Garrison Diversion Project appropriation in 1965. From that moment on, conservation organizations, the Audubon Society and the National Wildlife Federation, in particular, backed heavily by waterfowl hunting industries and downstream navigation interests, placed the Garrison Diversion Project under attack. The Project won its early court battles and construction began. Opposition intensified. Emissaries from downstream took up residence in North Dakota to spread poisonous attacks on Garrison Diversion Project by every device possible, from town meetings and talk shows to media articles, but construction continued. In desperation, the interests who wanted no water diverted from the Missouri River in North Dakota went into Canada and convinced a group of Canadians that if any Missouri River water got into the Red River of the North, Canada could expect dire consequences. Finally, the Garrison Diversion Project was driven to its knees when the International Joint Commission caved in to Canadian concerns.

The Garrison Diversion Project's McClusky Canal was well along, and the key Lonetree Reservoir was nearly complete when a surprise piece of legislation, introduced by the Audubon Society, appeared in the U.S. Senate, requiring a review of the entire Project by a ten-member commission. The legislation passed, and the commission, with two members from North Dakota, effectively killed chances of diverting Missouri River water out of its basin into the Red River of the North.

But life goes on. And life will go on for centuries. With burgeoning population and shifting demographics, potable water and agricultural water will become ever more precious. There will come a day when Missouri River water will be diverted to the Red River simply because a greater priority will develop. That day may be 50 years or 500 years from now, but it will come.

The Canadian objection that the Missouri River water might contain foreign fish species, biota, and fish pathogens that could infect the water of the Red River had to be addressed. We needed to know if the Canadian concerns were valid and if they could then be controlled or any possible losses mitigated. There was only one way to find the answers to the interbasin transfer of water concerns and that was through long-term scientific research.

Precise scientific research must stand alone above wishful thinking when studying the interbasin transfer of water. The argument cannot be won by pointing out that centuries ago, before the last ice age, the Missouri River flowed north to Hudson Bay; and, to this day, there are occasional flood situations at the extremes of the Red River where the waters of the two basins mingle. It makes little difference that millions of waterfowl have for centuries flown a few feet from the water and mud of the Missouri River basin to that of the Red River basin, and that aquatic animals have wandered back and forth across those few feet.

For decades, the North Dakota Game and Fish Department ignored any hazard of defiling Red River water when dumping truck loads of Missouri River basin water containing millions of fingerlings to be planted into the Red River basin lakes and streams. Never mind that, from the beginning of settlement, fishermen transported minnows, catches, and boats from one basin to the other.

Because of the possibility that foreign fish species, biota, and fish pathogens could cause significant economic loss to existing fisheries through interbasin transfer of water, it was recognized early on that reliable research had to be instituted. In 1986, as a member of the North Dakota State Water Commission, I went to Governor George Sinner and expressed my belief that a research program needed to be started. He gave me the green light. I began by inviting appropriate researchers to meet with me at North Dakota State University. Interest was apparent, but resources and funding were scarce. With assurance that some research would be undertaken, I requested the North Dakota State Water Commission to show state support by setting up a significant annual research grant. The Commission agreed.

Next, I sought out the Garrison Diversion Conservancy District board of directors. I pointed out that the only apparent way their eastern North Dakota county members would get Missouri River water was by overcoming Canadian fears of Red River contamination. The Conservancy board agreed to match the North Dakota Water Commission grant.

A research program was instituted. Dr. Jay Leitch, Associate Director of the North Dakota Water Resources Research Institute, agreed to direct the research. Gene Krenz of the North Dakota Water Commission staff was appointed to act as project coordinator at the state level. It was determined early on that the research work should enlist the Bureau of Reclamation and the Canadians in the studies, monitoring, and funding wherever possible. The North Dakota Game and Fish Department and the Health Department were expected to be involved.

I am pleased to see the research efforts on interbasin transfer of water focused in one publication. My hat is off to the dozens of people who quietly dedicate their life work to making living safer and more productive for coming generations. And, one thing we can be sure of is that much of this country's future depends on the wise use of an assumed water supply. In eastern North Dakota, water just might have to come from the Missouri River.

William L. Guy
Governor, North Dakota 1961-1973

About William L. Guy

At some point during each year from 1930 to 1940, the Red River of the North ceased to flow at Fargo, North Dakota. That and other experiences of the "dirty Thirties" left an indelible impression on young Bill Guy. As a youngster growing up in Amenia, a small eastern North Dakota village that was always short of water, he was acutely aware of the area's water problems.

In 1961, newly elected Governor of North Dakota Bill Guy was thrust into the politics of regional water development when his gubernatorial peers on the ten-state Missouri River States Committee of Governors elected him chairman for a two-year term. The committee served as the political arm of regional water resources development in the Missouri River basin. It was influential in persuading Congress to pass legislation establishing the National Water Resources Council in 1965, which encouraged river basin states to organize commissions to work on an equal footing with federal agencies in planning water projects that crossed state lines. Governor Guy was successful in organizing the Souris-Red-Rainy River Basin Commission with Minnesota and South Dakota in 1966. It was not until 1972, when Guy was able to overcome fears of a Missouri Valley TVA (Tennessee Valley Authority), that he was able to organize the ten-state Missouri River Basin Commission. President Reagan abolished the commission in 1981.

Governor Guy was a leader among the North Dakotans who finally persuaded Congress to reauthorize and appropriate funds for the start up of the long delayed Garrison Diversion Project in 1965. During Guy's 12 years in office (1961-1973), the Federal Fish and Wildlife Service's wetlands protection program began. More wetland acreage was acquired in North Dakota than in any other state. Serving as chairman of the North Dakota Water Commission, during his tenure as governor, gave him rare insight into the state's strengths and weaknesses in the way it managed its water resources. Since leaving the office of governor, Guy continues to promote water resources development and works closely with the North Dakota Water Users Association. When Governor George Sinner took office in 1985, Guy accepted a two-year term to return to the North Dakota Water Commission.

For six years, Guy was a water resources consultant to Basin Electric Cooperative, headquartered in Bismarck, North Dakota, which was builder and operator of the huge Missouri Basin Power Project (MBPP) electric generating plant in Wheatland, Wyoming. He worked on water rights for the cooperative's coal-fired electric power plants. This work ranged between Montana, Nebraska, North Dakota, South Dakota, and Wyoming. Guy also served as the MBPP representative on the three-person board of directors of the Platte River Whooping Crane Critical Habitat Maintenance Trust, headquartered in Grand Island, Nebraska.

Bill Guy's experience in water resources matters through his association with Congress, state and federal agencies, industry, municipalities, recreation, and agricultural groups gives him a broad background for his observations.

Foreword
Robert N. Clarkson

"Benefits for North Dakota, but not at Manitoba's expense!" Although the actions taken by Manitoba in its opposition to the construction of the Garrison Diversion Unit as originally designed may lead some North Dakotans to doubt this approach, this was the basic instruction under which the Government of Manitoba's Garrison Focus Office operated.

Manitoba became concerned with the effect that the irrigation of large areas of land within the Hudson Bay basin would have on the water quality of the Souris and Red Rivers flowing into Manitoba. This initial concern related more to the leaching of chemicals from the soil and the runoff of agricultural chemicals than the possible transfer of biota from the Missouri River basin to the Hudson Bay basin. This concern gave rise to the decision by the governments of Canada and the United States to refer the proposed project to the International Joint Commission (IJC) and the subsequent formation of the International Garrison Diversion Study Board. The IJC report in1977 confirmed Manitoba's concerns in respect to water quality and expressed even greater concern in respect to the matter of biota transfer between the two water basins. However, despite the expenditure of great sums of money and the best intentions, Garrison Diversion Unit, even as modified, presents an unacceptable risk of the introduction of unwanted foreign biota to the Hudson Bay drainage basin to the detriment of the people of Canada and to the general ecology of the region and beyond.

The Commission, therefore, concludes that, even if modified as described herein, the Garrison Diversion Unit will still cause adverse impacts in Canada. Only the extent of the impacts is in question. The Commission further concludes that while most of the impacts can be mitigated, those from the possible biota transfers are so threatening that the only acceptable policy at present is to delay construction of those features of the Garrison Diversion Unit which might result in such transfers.

Manitoba was now concerned not only with the effect that the project would have on the quality of the waters of the Souris and Red Rivers, but also with the risk that the possible introduction of foreign biota would present to Manitoba's commercial fishing industry, which annually generated $35 million. In 1981, the United States-Canada Consultative Group of American and Canadian federal officials and North Dakota and Manitoba officials was formed to review the concerns that Canada and Manitoba had advanced. The Consultative Group established a Garrison Joint Technical Committee with task forces on engineering and biology. These committees and the Consultative Group met on many occasions. In 1984, Canada urged that the United States undertake a study of alternatives to the proposed project features in the Hudson Bay drainage basin, and the United States noted that some studies were underway and agreed to support initiatives to broaden them.

When it became obvious that there was strong opposition to the Garrison Diversion Unit from the National Audubon Society and the National Wildlife Federation, as well as Canadian concerns, and the United States Congress was not prepared to continue to support the funding for the Garrison Diversion Unit as authorized, legislation was advanced establishing a Garrison Diversion Unit Commission. The Commission was directed to review the project and make recommendations to meet the contemporary needs of North Dakota and consider means of lessening the environmental impacts and concerns expressed by the Audubon Society, the Wildlife Federation, and Canada.

Manitoba's presentations to the Commission emphasized two matters: recognition of North Dakota's right to benefits from the Garrison Project and the need to ensure that the benefits which North Dakota received were not at the expense of Canada or Manitoba. While expressing its concern about the possible transfer of biota between the two drainage basins, Manitoba expressed its belief that the need for municipal, rural, and industrial water within North Dakota's portion of the Hudson Bay drainage basin could be satisfied without undue risk through the provision of Missouri River water properly treated before its transfer to the Hudson Bay drainage basin. The Garrison Diversion Unit Commission's 1984 report again recognized Canada's legitimate concerns in respect to the danger of biota transfer.

The Garrison Diversion Unit Reformulation Act of 1986 adopted the major recommendations of the Garrison Diversion Unit Commission. The development and approval of the act represented a very significant degree of compromise by all parties involved. The redesigned project, while meeting the contemporary water needs of North Dakota, significantly reduced the impact of the project on the environment and provided a reasonable level of protection against the transfer of biota between the two water basins.

The United States-Canada Consultative Group and its Engineering and Biology Task Forces commenced meeting again in 1989 to further assess the reformulated project. The Biology Task Force undertook a study of the present distribution of fish species and pathogens of concern and the relative level of risk of transfer of biota from the various components of the reformulated project. It pointed out in its report of November 1990 that the concerns expressed are real and legitimate. It matters not to a receiving system whether an introduction has been deliberate so as to enhance a recreational fishery, to establish aquaculture ventures in a new area to develop potentially profitable markets, for the biological control of plants and insects, to fill 'vacant' niches in man-made lakes and reservoirs, for ornamental purposes and the aquarium trade, or inadvertent, such as a result of water diversion schemes or the discharge of ballast water from foreign freighters; or unauthorized as a result of intentional or accidental bait-bucket transfers; the end result in all cases will be the same.

Since introduced plants, animals, and pathogens have no respect for state, provincial, or international boundaries, they expand within areas with suitable climate and favourable physical, chemical, and biological conditions. Downstream users of aquatic resources often find themselves to be the unknowing and/or unwilling recipients of the non-native species introduced upstream. Once an exotic or non-native species becomes established in a new habitat, it is usually impossible or impractical to eradicate it; it becomes, for better or worse, a permanent addition to the flora or fauna.

There are many examples of the effect of the introduction by man, both intentional and accidental, of foreign species of fish into new host water basins and the dramatic effect such introductions have had on the receiving water basin—the carp into the New England States and its eventual migration to most of the water basins of North America east of the Rocky Mountains; the alewife, the sea lamprey, and zebra mussels into the Great Lakes. Each has had adverse effects on the native inhabitants of the basin into which it was introduced.

This study addresses many of the scientific and technical questions in regard to the risk and effects of biota transfer between water basins. Several of the studies that were commissioned were undertaken in Canada by Canadian scientists and have been carried out with the cooperation of the Department of Zoology, University of Manitoba. However, the commercial fishermen on Lake Winnipeg and sports fishermen on the Red River and its lakes and tributaries will continue to be concerned about any changes that may adversely affect their harvest.

Dr. Rodney Sayler of the Institute for Ecological Studies, University of North Dakota, said it well when he stated in his Status Report and Analysis of Biota Transfer issues:[1] From an ecological perspective, it is unlikely that future objections to interbasin water transfer will change even if selected fish species and other organisms from the Missouri River Basin are found to occur in the Hudson Bay Basin...unforseen biota transfer problems (e.g., new pathogens/fish introductions) also could develop in the future. Ecological data are inadequate to predict potential impacts of biota transfer with any degree of certainty.

Robert N. Clarkson
Coordinator, Garrison Focus Office

Province of Manitoba, 1983-1994

[1]Sayler, R.D. 1990. Fish transfers between the Missouri River and Hudson Bay basins: a status report and analysis of biota transfer issues. In Leitch, J.A. and D.J. Christensen (eds.) *1990 Interbasin Biota Transfer Study Program Proceedings, North Dakota Water Quality Symposium.* North Dakota Water Resources Research Institute, North Dakota State University, Fargo.

About Robert N. Clarkson

Bob Clarkson was born at Grand Prairie, in the Peace River country of northern Alberta, in 1928. His family moved to southern Saskatchewan in 1941, where Bob finished high school and commenced work as a banker. Following transfers by the bank to Winnipeg, northern Ontario and back to Winnipeg, Bob resigned from his position as manager of a branch bank in Winnepeg in 1963 to take a position with the civil service of the Government of Manitoba.

Administrative responsibilities with the Department of Municipal Affairs were followed by secretarial and financial responsibilities for the Government of Manitoba's public housing corporation. In 1979, Clarkson left his position as General Manager of The Manitoba Housing and Renewal Corporation to assume responsibility for research and coordination of a review of real property assessment in the Province of Manitoba. The report of the Manitoba Assessment Review Committee, "A Fair Way to Share," was published in 1982.

Upon completion of this report, Bob was asked to take responsibility for the coordination of special programs within Manitoba Natural Resources. His responsibilities included the negotiation with towns and villages of the Red River Valley for the upgrading of ring dykes, which had been constructed following the flooding of 1950 to protect these communities (including the agreement with the U.S. Corps of Engineers for the joint protection of Noyes, Minnesota, and Emerson, Manitoba); the analysis of residential problems within Provincial Parks; the coordination of the input of the staff of Natural Resources in the resolution of claims associated with the construction of hydro dams on the Churchill and Nelson River systems.

In 1983, Clarkson was asked to also assume responsibility for the Manitoba Garrison Focus Office, an office established by the Government of Manitoba to coordinate and bring forward its concerns and the concerns of non-government organizations with respect to the Garrison Diversion Project in North Dakota.

As coordinator of the Garrison Focus Office, Clarkson was responsible for the dissemination of information about the Garrison project to all members of the Manitoba Legislature and the organization of lobbying efforts in Washington by political members of both the government and the opposition at the provincial and federal level. His responsibilities also included the coordination of efforts and positions with the National Wildlife Federation and the National Audubon Society, and liaison with the Canadian Department of External Affairs, Environment Canada, and the Canadian Embassy in Washington, D.C.

Clarkson served as Manitoba's chief representative to the United States–Canada Garrison Diversion Consulative Group and on its Engineering

Task Force which attempted to resolve Manitoba and Canada's concerns respecting construction plans for the Garrison Diversion Project.

Additional international responsibilities included representing Manitoba in negotiations of the United States–Canada agreement for the construction of the Rafferty-Alameda dams on the Souris River in Saskatchewan and its effect on water quality and flows to North Dakota and from North Dakota to Manitoba.

Clarkson retired from the civil service of the Province of Manitoba in 1994; however, he continues to be engaged as a consultant to Manitoba on matters associated with the settlement of various land claims of Manitoba's native population. Bob Clarkson's strength in negotiations and resolutions of problems comes from his belief that there can be no resolution of a problem unless both sides fully understand the position and concerns of the other side.

Preface
Jay A. Leitch

The Missouri River, North America's longest river at 2,540 miles with a drainage basin of 529,000 square miles, empties into the Mississippi River near St. Louis, Missouri, with an average discharge of about 76,000 cfs. Across the continental divide on the North Dakota prairies, the Red River of the North meanders nearly straight north for over 500 miles, emptying into Lake Winnipeg and, eventually, Hudson Bay. A component of the 1940s Pick-Sloan Plan to develop the Missouri River was to divert water from the Missouri River to irrigate one million acres of arid land in western North Dakota. In 1965, the United States Congress authorized irrigation of 250,000 acres lying, across the continental divide, within the Red River of the North drainage basin. In October 1973, the Canadian government, supported by the Manitoba Naturalists Society, the Canadian House of Commons, and the Canadian Department of External Affairs, requested a moratorium on the project to protect Canadian interests under the 1909 Boundary Waters Treaty, which prohibits the pollution of shared watersheds. The government of the United States recognized its obligations under the treaty and halted construction on the project in February 1974, until Canadian objections could be addressed (Kelly and Leitch 1990). Among other problems, there was a potential for undesirable aquatic species to be transported across basin boundaries.

In 1986, the United States government mandated that an international study team assess the biota transfer problem before any diversion water crossed the continental divide. (Garrison Diversion Reformulation Act of 1986, P.L. 98-360). In 1987, proponents of the Garrison Diversion Unit (GDU) turned to science to help ameliorate domestic and international concerns about inadvertent transfer of aquatic biota. After a scoping study to determine issues, funds were made available to begin multidisciplinary studies in May 1988. A team of scientists was selected to identify specific, researchable topics; to develop requests for proposals; and to select principal investigators to carry out research. From 1987 through 1995, 20 studies, involving scientists from both the United States and Canada, representing several scientific disciplines were funded; all but two of the studies were completed by early 1996.

A review of interbasin transfers around the world found that biota transfer concerns have only recently received much attention (Padmanabhan et al. 1990). There are many unresolved issues with respect to biota transfer as a result of past, present, and planned interbasin water transfers worldwide. Cases providing the most information were in the United States, Canada, South Africa, India, China, and the former Soviet Union, although interbasin water transfers have occurred worldwide. Interbasin transfer proposals have recently encountered a host of political, socioeconomic, engineering, and legal problems. Factors responsible for these concerns stem from strong opposition expressed by exporting basins, economic feasibility questions created by rising costs, and environmental considerations which have brought about a more cautious approach (Greer 1983).

Scoping study

A scoping study to identify and categorize Canadian concerns and issues regarding GDU (Leitch and Grosz 1988) identified five dimensions of biota transfer that needed further study:

- fish screens,
- underground return flows,

- distribution/life cycles of specific biota and pathogens,
- biota transfer via municipal and industrial uses of water, and
- biota transfer via operational failures.

The North Dakota Water Resources Research Institute proposed, and was selected, to be overall study project manager. It would manage a study program to identify research that:

- has the characteristics of good science,
- responds to issues pertinent to all parts of the Conservancy District,
- resolves short run issues,
- resolves issues that require a longer term, and
- shows high likelihood of resolution with a high probability of acceptance by the scientific community.

A Technical Advisory Team (TAT) was formed to assist in identification of projects and selection of principal investigators. The TAT included members of the scientific community from North Dakota, Minnesota, and Manitoba. Disciplines represented on the committee included Biology, Systems Ecology, Civil Engineering, Zoology, Chemistry, and Resource Economics.

Canadians were concerned specifically about four fish species: Utah chub *Gila atraria*, shortnose gar *Lepisosteus platostomus*, gizzard shad *Dorosoma cepedianum*, and rainbow smelt *Osmerus mordax*. Some fish pathogens were also identified (IHN, infectious hemapoietic necrosis; entenic red mouth, *Yersinia ruckeri*; and *Polypodium hydriforme*) as potentially harmful to Hudson Bay basin ecosystems. However, no human pathogens, aquatic plants, or other flora or fauna were identified early on in the controversy over biota transfer.

Four general areas of concern were identified as researchable and meeting the five characteristics mentioned. First was how to prevent biota from being transferred. Second was to assess whether the aquatic species of concern already inhabited the Hudson Bay basin. Third was to assess whether or not they would survive if transported across basin boundaries. The fourth general concern was to assess the impact on the receiving ecosystems.

Preventing biota transfer

Direct filtration was explored first as a means of preventing biota transfer (Turner and Hefta 1990). Direct filtration is a water treatment process that does not use sedimentation prior to filtration and generally involves rapid mix and flocculation immediately prior to filtration and disinfection following filtration. While direct filtration systems have been shown effective in removing human pathogens, they have not been used for removal of fish pathogens. A dual media system with a drain to a large sump where organisms are destroyed by chlorination was designed and tested at bench scale. Results indicated that filters should remove all fish, fish eggs, and fish larvae from the water. Additionally, the processed water would meet United States Environmental Protection Agency standards for public drinking water.

Application of ozone as a treatment method for prevention of interbasin transfer of aquatic biota was studied as a way to further ensure prevention of biota movement. Samples of actual diversion water were inoculated with fecal coliform organisms (an indicator organism) and treated at various ozone CT (concentration x time) values to evaluate disinfection efficiency using bench scale batch and continuous flow equipment. Results indicated extremely high levels of disinfection.

Species ranges

An initial assessment of the range distribution of the undesirable fish species found they had not changed much since the concerns were initially raised more than ten years earlier, although rainbow smelt and gizzard shad were found in some waters where they had not been reported earlier. In fact, rainbow smelt were found in the Hudson Bay basin in September 1990, thus eliminating them as a species of concern. Also, introduction of the zander *Sizostedion lucioperca*, a European species, by the North Dakota Game and Fish Department raised some concerns (Courtenay and Robins 1989). The zander program has since been dropped from consideration by the North Dakota Game and Fish Department.

Although limnological and fisheries research has been carried out on Lake Winnipeg for at least 60 years, no attempt had been made to collate or summarize the information gained into one document. After finding little information available on the limnology and fisheries of Lake Winnipeg, an intensive three-year project was developed. This data will be useful to provide insights into the responses of basin biotic communities to alterations in component parts produced by invading species and climate change.

A study of the distribution and dispersal of fishes in the Red River basin was conducted to bring together existing references, as well as to conduct field work to fill in voids. Emphasis was placed on the success of white bass *Morone chrysops* and gizzard shad as colonizers, since they are found basin-wide. Results will be used primarily to provide baseline information.

There are a variety of pathways for aquatic species range expansion, including both natural and anthropogenic. Natural pathways include connections between basins at times of high water, animal transport, and extraordinary weather events (e.g., tornadoes, hurricanes). Introductions by humans also can result from a variety of activities, including escapes from aquaculture, aquarium releases, stocking activities, ballast release, and angler escape or release. The undesirable species may be serving as bait, an aquaculture species, or an aquarium species, or it may be hitchhiking in water used to transport desirable species. A study was carried out to specifically address the potential for bait bucket transfer—the probability that anglers using live bait will contribute to biota transfer.

Authors of the bait bucket study concluded that while the potential for bait bucket transfer is low for a single angler for a single angling day, the cumulative effect of 19 million angling days, over the course of a year in the study area, brings the annual probability close to (but, of course, never reaching) 1.0 (Ludwig and Leitch 1996). The high probability can be attributed to increased angler mobility, baitfish maintenance technology, and the long distances traveled by bait wholesalers. The implication is that expensive efforts to prevent biota transfer as a result of an interbasin water transfer plan may be in vain unless the potential for bait bucket transfer is substantially reduced.

Invasive species survivability

While the survivability of invasive species was identified as an area where science could provide useful information, specific studies to address the issue were never carried out. However, species indigenous to the Missouri River would likely survive in adjacent portions of the Hudson Bay basin, especially if the hypothesis about glacial disruptions of ranges (Missouri River species are returning to niches previously vacated during the glacial period) cannot be rejected.

The usefulness of an ecosystem model became apparent early in the course of the overall study. However, just as information was becoming available (i.e., stonecat model, Lake Winnipeg assessment, Red River fish inventory) to begin to construct a model, project sponsors chose to reduce the scope of investigations to prevention only, thus, eliminating the need for a model to assess impacts of invasive species.

Invasive species ecological impact

Invasive species could impact native ecosystems in several ways, including serving as a host for parasites or pathogens, or competing for space, food, or other niche aspects. The stonecat *Noturus flavus* was studied as a model for the dispersal of invasive species (Stewart and McCulloch 1990; McCulloch 1994). Stonecat were found to share habitat with an indigenous species, longnose dace *Rhinichthys cataractae*, leading Stewart to eventually conclude the stonecat might have been displaced by glaciers and are merely reinhabiting the Hudson Bay basin.

Research carried out on parasites and pathogens of paddlefish *Polyodon spathula* and lake sturgeon *Acipenser fulvescens* concluded that some parasites and pathogens were already present in both basins. This research was both the most "basic" and the most expensive due to a paucity of baseline data in this area. Results contributed far more to science than to the resolution of the issues primarily addressed by the overall study.

Observations

Two types of observations can be drawn from this study. First, several observations about organization of applied, multidisciplinary research can be made. Second, the role of science in political-environmental decision making can be clarified.

Research organization

While the study was initiated and largely funded by proponents of interbasin water transfer, its organization insulated scientists from study sponsors' specific interests. Researchers were strongly encouraged to publish in peer-reviewed outlets, lending credibility to their results. A master's thesis was written to document organization of the overall interbasin aquatic biota research program (Christensen 1993). One conclusion of the thesis was that it takes several types of scientists to effectively carry out such a study. Intradisciplinary scientists are necessary for in-depth studies in a single discipline. Interdisciplinary scientists are needed to bring the components of similar disciplines together. Multidisciplinary scientists are required to make the connections among several unrelated disciplines. Finally, a generalist with the help of an advisory board is needed to manage all the activities, to ensure funds are available when needed by scientists, to integrate sponsor expectations with research results, and to communicate needs to scientists and results to lay audiences.

Role of science

While it often seems that "someone must have done that already," more often than not there are gaps in the science and in the baseline data necessary to assess environmental changes adequately enough for policy making. Roughly $US1 million has been spent in direct costs and as much as $US1 million in in-kind support to support science-based applied research to help resolve what was largely a politically-based problem. While the project resolved some issues, as does any scientific investigation, it also raised additional issues. In spite of the special interest reasons for the study, it nonetheless did advance science along a number of fronts, including the study of interbasin transfer of aquatic species.

Science can provide a less subjective, broader, defensible base for decision making. It can offer information that is held to be true "beyond a reasonable doubt," but doubt is never entirely and absolutely absent. It is possible, and in fact necessary, to use science to help solve environmental management problems. However, employing science does not guarantee that solutions will be reached. The incorporation of science into the decision-making process should follow a rational process and should be accompanied by specific expectations on the part of scientists and decision makers.

It is not possible to design a water transport system that would be fail safe; however, the probability of transfer could be reduced to an extremely small number through the use of physical and chemical treatments. Sand filters and ozone treatment effectively remove almost all traces of aquatic biota from treated water. [Given the cost of these systems and the probability that aquatic biota could be transferred via other mechanisms raises questions about their efficiency.] Many other natural and anthropogenic pathways were identified for aquatic biota transfer between the two watersheds. Anglers' bait buckets had a probability of near 1.0 to transport some species over the course of a year due to the extremely high number of events (angler days) in the study area. Naturally, high water, bird and mammal transport, hitchhiking with hobby and commercial aquaculture species, and extraordinary meteorological events could each be responsible for aquatic biota transport.

At least one fish species of concern to Canadians, rainbow smelt, was found to already inhabit the Hudson Bay basin suggesting the probability that other species could either already be there or be likely to get there via other means. There is a possibility that some species may be returning to former habitats disrupted by glacial activity about 10,000 years earlier, that they were part of the Hudson Bay fauna at one time and will return to a vacant or nearly vacant niche.

In spite of an extensive effort to ameliorate biota transfer concerns, Canadian concerns have not appeared to lessen. While very recent accounts of continued concerns are largely unpublished, as recent as 1992 in a presentation in North Dakota, Clarkson (1992) identified as many as ten additional "unwanted" fish species, several other problems with Garrison Diversion, and reiterated provisions of the 1909 Boundary Waters Treaty and the Garrison Diversion Unit Reformulation Act of 1986. In short, due both to the uncertainty of science and the vagaries of politics, it appears that science will never be able to fully satisfy the concerns of Canadians regarding the potential for interbasin transfer of aquatic biota. Given that outlook, the study sponsors choose to concentrate their research efforts beyond 1996 strictly on prevention.

References

Christensen, D. J. 1993. *The Sociology of Multidisciplinary Research: A Case Study*. M.S. thesis. Fargo: North Dakota State University.

Clarkson, R.N. 1992. *Canadian Concerns Regarding the Garrison Diversion Unit.* Third Biennial North Dakota Water Quality Symposium, Bismarck, North Dakota.

Courtenay, W.R., Jr. and C.R. Robins. 1989. Fish Introductions: good management, mismanagement, or no management? *Aquatic Sciences* 1:159-172.

Greer, C. 1983. The Texas water system: implications for environmental assessment in planning for interbasin water transfers. In *Long-Distance Water Transfer: A Chinese Case Study and International Experiences*, 77-90, ed. Biswas, A.K. Dublin: Tycooly International Publishing Ltd.

Kelly, P.E. and J.A. Leitch. 1990. History of Missouri River diversion and biota transfer. In *Proceedings of the North Dakota Water Quality Symposium, March 20-21, 1990*, 234-251. Fargo: North Dakota State University, Extension Service.

Leitch, J.A. and K.L. Grosz. 1988. *Identification and Analysis of Canadian Concerns Regarding the Garrison Diversion Unit in North Dakota.* Fargo: North Dakota Water Resources Research Institute.

Ludwig, H., Jr. and J.A. Leitch. 1996. Interbasin transfer of aquatic biota via anglers' bait buckets. *Fisheries* 21(7): 14-18.

McCulloch, B.R. 1994. *Dispersal of the Stonecat (Noturus flavus) in Manitoba and Its Interactions with Resident Fish Species.* M.S. thesis. Winnipeg: University of Manitoba.

Padmanabhan, G., K. Jensen, and J.A. Leitch. 1990. A review of interbasin water transfers with specific attention to biota. In *Proceedings of the Symposium on International and Transboundary Water Resources Issues*, 93-99, ed. Fitsgibbon, J.E. Bethesda, Maryland: American Water Resources Association.

Stewart, K.W. and B. McCulloch. 1990. The stonecat, *Noturus flavus*, in the Red River-Assiniboine River watershed: progress report on an invading fish species. In *Proceedings of the North Dakota Water Quality Symposium,* 264-280. Fargo: North Dakota State University, Extension Service.

Turner, C.D. and M. Hefta. 1990. Removal of biota from inter-basin transfer water. In *Proceedings of the Symposium on International and Transboundary Water Resources Issues*, 129-138. Bethesda, Maryland: American Water Resources Association.

CHAPTER 1
Introduction
David R. Givers and Jay A. Leitch

Science is often called upon to help resolve problems affecting society. However, to avert any potential misunderstanding about the ability of science alone to resolve issues, the role of science in society needs to be clarified.

Barriers to social problem solving may occur when contentious parties do not agree on problem definition, including whether a problem exists, what the nature of the problem is, and whether the problem, perceived or real, actually matters (Trudgill 1990). In the expectation of clarifying and resolving controversial issues, policymakers may turn to science to help identify and define the problem(s).

There are two ways in which science may serve society and policymakers. First, science may produce specific, objective, or factual information on technical problems (a descriptive process). Second, scientific techniques can be used to define problems and provide a framework or a logical and systematic method for choosing solutions, or furnish a method for weighting choices (a decision-making process).

Science provides methods for problem identification and definition, while supplying the best source of objective data to describe systems and conditions. Science may not provide a single definitive outcome value, nor a single best choice, but rather, it can provide ranges of a single outcome or probabilistic confidence intervals and/or several alternative choices. The selection of the best choice (policy decisions) is usually reserved to the policy arena and influenced by more than just science.

Scientists are often asked to serve as advisors to governmental and non-governmental organizations. Science advisors are asked to interpret science, to render advice, and perhaps to shape policy objectives. Policymakers may mistakenly assume that the scientist can render opinions rationally and dispassionately, but there is no evidence that this is so (Trudgill 1990; Bundy 1963). Studies of scientific advising leave in tatters the notion that it is possible, in practice, to restrict the advisory process to technical issues or that the subjective values are irrelevant to decision making (Jasanoff 1990, p. 230). This does not imply that scientists should be categorically prohibited from rendering advice and expert opinion any more than business managers, farmers, or ordained ministers be intentionally excluded from participating in policy formulation. Rather, scientists, when acting in advisory roles, are often not scientists, but educated citizens. Knowledge required for advising on social policy issues is distinctly different from the knowledge of most scientific disciplines, and disciplinary knowledge does not automatically confer wisdom in social problem solving.

Limits of science

The building block of all science-based decision making is research (fig. 1.1). Research spans a continuum from purely basic to applied. Basic is also called disciplinary, while applied is further divided into subject matter and problem solving. Basic science is conducted primarily by academics. Pracademics work to move the fundamental research closer to practical, or more commonly useful, applications of the underlying science. This is an important distinction. Basic science often does not have immediate applicability either for commercialization of goods and services or for policy

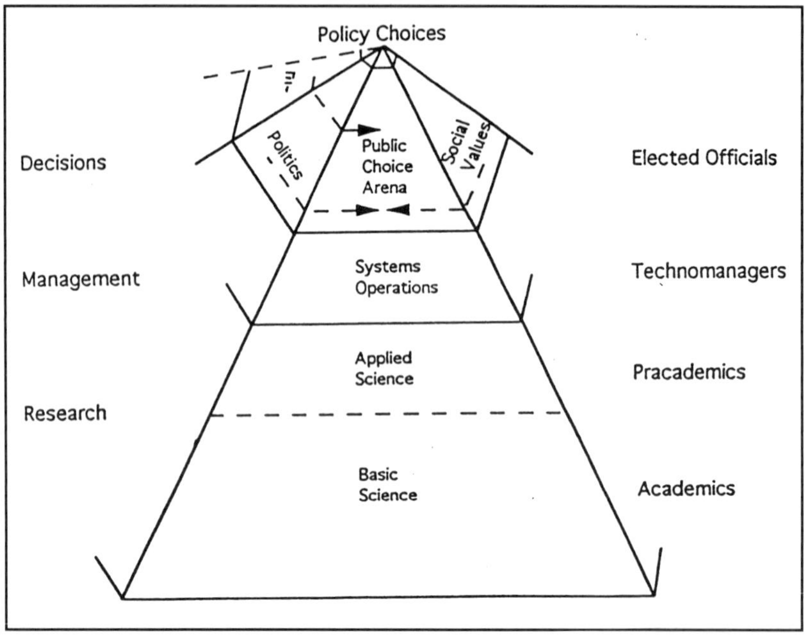

Figure 1.1. Science-Based Decision Making.

formulation and administration. Pracademic applications can help move basic research into product development and the social policy arena.

Science can provide techniques for organizing problem-solving exercises and methods for testing the validity of hypotheses and the credibility of data. Science can also be used to narrow the range of policy options available "beyond a reasonable doubt," even though "reasonable" may be subject to wide variation. The role of science in social problem solving is vital but limited.

A hierarchical relationship of science to social decision making has evolved into democratically-organized societies. Science is only one component of society, so it cannot organize or direct social policy. It is not the fourth branch of government. Final decisions, choosing among the range of options defined or developed through research and science-based recommendations, rest with policy managers and elected officials.

Some scientific disciplines, such as the natural and physical sciences, have traditionally carried out their efforts apart from the human element. However, neither the environmental issues of today nor today's socioeconomic-political environment can be understood or managed in society's best interest apart from the other. That is why the study of the interbasin transfer of waterborne biota should include scientists from many disciplines—from biology to economics.

Ecosystem science and environmental decision making

Ecosystem science (ecology) is still in its infancy. The earliest recognized use of the term ecosystem was in 1869 by German biologist, Ernst Haeckel. Specific discoveries, writings, and theoretical applications followed Haeckel's use of the term. An increase in the number of scientists pursuing ecological studies in North America led to the founding of the Ecological Society of America in 1914.

Scientific knowledge is cumulative and dynamic, subject to change as each generation makes its contributions. Early students of ecology, such as George Perkins Marsh (1874), established a framework of thinking which focused on the interaction of natural resource systems and the relation of humans to the natural environment. The interconnection of humanity and the ecosystem remains, but our knowledge of specifics has obviously increased, resulting in modifications to ecological theory.

Ecological knowledge has expanded through data collection, hypothesis testing, modeling, interpretation, and analysis. Science typically relies on modeling to test theory and to make predictions. Models are simplifications of complex real systems. They can provide a generalized, but accurate, description of the behavior of systems or their components. The purpose of using a model is to develop theory which has descriptive or predictive value.

Ecosystem simulation or modeling, in its early stage of development, came of age only after the proliferation of computer-based technology. Models incorporating important environmental variables allow the variables to interact in unison as they do in real systems, serving to provide

- generalizable conclusions from specificity,
- precise assessment of specific components of the system, and
- realistic representation of existing conditions.

No existing ecological model can provide all three simultaneously (Swartzman and Kaluzny 1987, Levins 1966).

In fact, ecosystem modeling is only one step in the natural resources management problem-solving process and it is not even the first step (Chechile 1991, Trudgill 1990, National Research Council 1986). Given the comparatively recent contributions of ecological science to the understanding of ecological structure and functions, and given the limitations of models describing ecosystems, neither scientists nor decision makers should promise more, or expect more, than science can deliver about natural resources management issues. Society, nonetheless, cannot wait for scientists to refine models and theories — decisions must and will be made. Not making a decision is a decision to do nothing that has consequences just like a proactive decision.

Scientific study of biota transfer issues in the interbasin transfer of water

A scientific study is initiated in the belief that science can help provide better information for making decisions. Some people believe though that science is being ignored, succumbing to social and political pressures. At the same time, there are as many people who are convinced that social issues are being ignored or that legitimate political concerns are not given proper weight. This further accentuates the need for a well-developed, multidisciplinary study that considers as many relevant issues and perspectives as resources permit.

The North Dakota Governor's Oversight Committee identified potential aquatic biota transfer from the Missouri River drainage basin to the Hudson Bay drainage basin as the main concern in an ongoing interbasin water transfer controversy between North Dakota and Canada. Several researchable "scientific" issues were identified by the Interbasin Biota Transfer Study Program Technical Advisory Team (IBTSP-TAT) including:

- the efficacy of fish screens and filtration technology;
- the role, if any, of underground return flows;
- the distribution and life cycles of specific biota and pathogens identified as important to Canadian concerns; and
- the transfer potential via municipal and industrial water and from failure of Garrison Diversion Unit (GDU) operational systems (Leitch and Grosz 1988).

Scientists were provided funds in 1988 to begin working specifically to generate information to resolve these issues. TAT issued five rounds of Requests for Proposals (RFPs) over an eight-year period, inviting scientists in Canada and the United States to conduct research on these issues.

For example, fisheries scientists, making routine observations had noted that rainbow smelt *Osmerus mordax* were not present in the Hudson Bay drainage basin. Consequently, smelt was identified as a species with potential to damage or disrupt Canadian fisheries if Garrison Diversion transfer occurred (IJC 1977).

Given the lack of evidence to suggest smelt were present in the basin, science could not predict they would be observed in Lake Winnipeg on September 26, 1990, as was the case (Campbell et al. 1991). Even with smelt known to be present in Lake Winnipeg, scientists cannot reliably predict that a viable or competitive population will result, though that is one possible outcome. Smelt may fail to reproduce or extend its range in this new ecosystem, or it may invade and survive with no net change on the existing fishery (Moyle et al. 1986). Scientists will eventually be able to infer an outcome and perhaps, through expert judgement, assign a probability or likelihood of the impact within a narrow range of outcomes. Some of the limits of science are obvious in this example. Finding answers to researchable issues may not necessarily provide an indisputable base from which decision makers can formulate policy. However, science can narrow the range of choice within which decisions can be made.

Another example of a researchable concern is the effectiveness of mechanical filters and screens on biota transfer. During laboratory and bench testing, Turner and Hefta (1990) tested the hypothesis that direct filtration combined with disinfection removes pathogens and concluded that pilot-scale modeling was warranted. Follow-up pilot-scale research continued in RFP rounds four and five. Scale-up of equipment and economic feasibility would be the next phase, assuming the testable hypothesis is not subsequently rejected. Research, based on the use of models to replicate real world conditions, holds the potential for understanding biota transfer problems and identifying potential solutions.

Conclusion: role of science in contemporary problem solving

Science has much to offer, but science alone cannot resolve all the issues involved in interbasin biota transfer. Science can provide a less subjective, more comprehensive, defensible base for decision making. It can present information (descriptive or interpretive) that is held to be true "beyond a reasonable doubt." Yet doubt is never entirely and absolutely absent. "The role of the scientific method has more to do with correcting false ideas than establishing probable truth" (Armstrong and Botzler 1993, p. 2).

According to Leitch and Givers (1992), natural resources management policy making rests firmly upon a three-legged stool comprised of science and technology, economics, and politics (fig. 1.2). All three are equally important and necessary cornerstones to policy making. Technical know-how, supplied by science, is necessary to implement sound management goals (the knowledge facet of policy), and policy initiatives must be economically viable to be sustainable. The third leg is comprised of the political decision-making process where policymakers bring together all the facets of social choices regarding natural resources management.

It is possible, and in fact necessary, to use science to help solve environmental management problems. The incorporation of science into the decision-making process should follow a rational process and should be accompanied by specific expectations on the part of scientists and decision makers.

References

Armstrong, S. J. and R.G. Botzler. 1993. *Environmental Ethics: Divergence and Convergence.* New York: McGraw-Hill, Inc.

Bundy, M. 1963. The scientist and national policy. *Science* 139:805-809.

Campbell, K.B., A.J. Derksen, R. A. Remnant, and K. W. Stewart. 1991. First specimens of the rainbow smelt, *Osmerus mordax*, from Lake Winnipeg, Manitoba. *The Canadian Field Naturalist* 105:568-570.

Chechile, R. A. 1991. Introduction to environmental decision making. In *Environmental Decision Making: A Multidisciplinary Approach*, eds. Chechile, R.A., and S. Carlisle. New York: Van Nostrand Reinhold.

Figure 1.2, Natural resources management policy legs

International Joint Commission. 1977. *Transboundary Implications of the Garrison Diversion Unit.* Toronto, Canada.

Jasanoff, S. 1990. *The Fifth Branch: Science Advisers as Policy Makers.* Cambridge: Harvard University Press.

Leitch, J. A. and D. R. Givers. 1992. Water quality research and management challenges. *Proceedings: North Dakota Water Quality Symposium.* Fargo: North Dakota State University, Extension Service.

Leitch, J. A. and K. L. Grosz. 1988. *Identification and Analysis of Canadian Concerns Regarding the Garrison Diversion Unit in North Dakota.* Fargo: North Dakota State University, Water Resources Research Institute.

Levins, R. 1966. The strategy of model building in population biology. *American Scientist* 54(4):421-431.

Marsh, G. P. 1874. *Man and Nature*, 3d ed., ed. Lowenthal, D. Cambridge: The Belknap Press of the Harvard University Press.

Moyle, P. B., H. W. Li, and B. Barton. 1986. The Frankenstein effect: Impact of introduced fishes on native fishes of North America. In *The Role of Fish Culture in Fisheries Management*, ed. Stroud, R.H. 415-426. Bethesda, MD: American Fisheries Society.

National Research Council. Commission on Life Sciences, Committee on the Application of Ecological Theory to Environmental Problems. 1986. *Ecological Knowledge and Environmental Problem Solving.* Washington, D.C.: National Academy Press.

Swartzman, G. L. and S. P. Kaluzny. 1987. *Ecological Simulation Primer.* New York: MacMillan Publishing Company.

Trudgill, S. 1990. *Barriers to a Better Environment: What Stops Us Solving Environmental Problems?* London: Belhaven Press.

Turner, C. D. and M. J. Hefta. 1990. *Direct Filtration of Garrison Diversion Water*, Presented to North Dakota Water Resources Research Institute, Grand Forks: University of North Dakota, Department of Civil Engineering, Energy and Environmental Research Center.

CHAPTER 2
History of Garrison Diversion

Paul E. Kelly, Jay A. Leitch, Gene Krenz, and Mariah J. Tenamoc

Executive summary

The idea of diverting water from the Missouri River for economic and domestic purposes has been around for as long as people have resided in North Dakota. Hard times in the 1930s forced state leaders to reevaluate their relationship with the mighty river. Throughout the years, there have been many plans to ensure that the "dirty Thirties" would never happen again. Garrison Diversion has been seen historically as the panacea for most of North Dakota's problems. Yet the ideas and plans for diversion in themselves have created problems early leaders did not foresee.

The importance of this study indicates that any large-scale project that affects society the way this one has needs to be considered carefully. The input from the diverse disciplines involved with this study bears this out. This study's greatest strength is the involvement of many different scientists, government leaders, and concerned citizens to reach some kind of consensus. This study also reveals that Garrison Diversion still has a long way to go, if it goes at all.

Introduction

The United States Congress authorized the first stage of the Garrison Diversion Unit (GDU) in North Dakota in 1965 to lift Missouri River water from behind Garrison Dam to

- irrigate 250,000 acres of land,
- provide municipal and industrial water supply for fifteen towns and cities,
- provide water for twenty-four areas (146,520 acres) for fish and wildlife conservation, and
- provide water for recreational development at seven major water impoundments.

Missouri River water would be conveyed from Lake Sakakawea to Lake Audubon, into the McClusky Canal, and fill the Lonetree Reservoir, behind Lonetree Dam, near Harvey. From there, the water would enter the New Rockford Canal, which would feed Devils Lake and Stump Lake. Return flows would enter the James and Sheyenne Rivers. The Velva Canal would transfer water north from the Lonetree Reservoir, north into the Souris River for irrigation and municipal water supplies (fig. 2.1).

In October 1973, the Canadian government, with the support of the Manitoba Naturalists Society, the Canadian House of Commons, and the Canadian Department of External Affairs, requested a moratorium on the project to protect Canadian interests under the Boundary Waters Treaty of 1909, which prohibits the pollution of shared watersheds. In February 1974, the government of the United States recognized its obligations under the treaty and halted construction on the project until Canadian objections could be addressed (International Joint Commission 1976). To understand why the Canadian government sought a moratorium on the Garrison Diversion Project, one must first look at the GDU itself, how it came about, and the changes in the unit that led to Canadian objections (fig.2.2).

The beginning years (1883-1914)

North Dakotans look to the Missouri River as a source of navigation, irrigation, and municipal water supplies. As early as the 1800s, members of the American Fur Company entered the Louisiana Territory to exploit the abundance of fur-bearing animals and used the Missouri River to transport

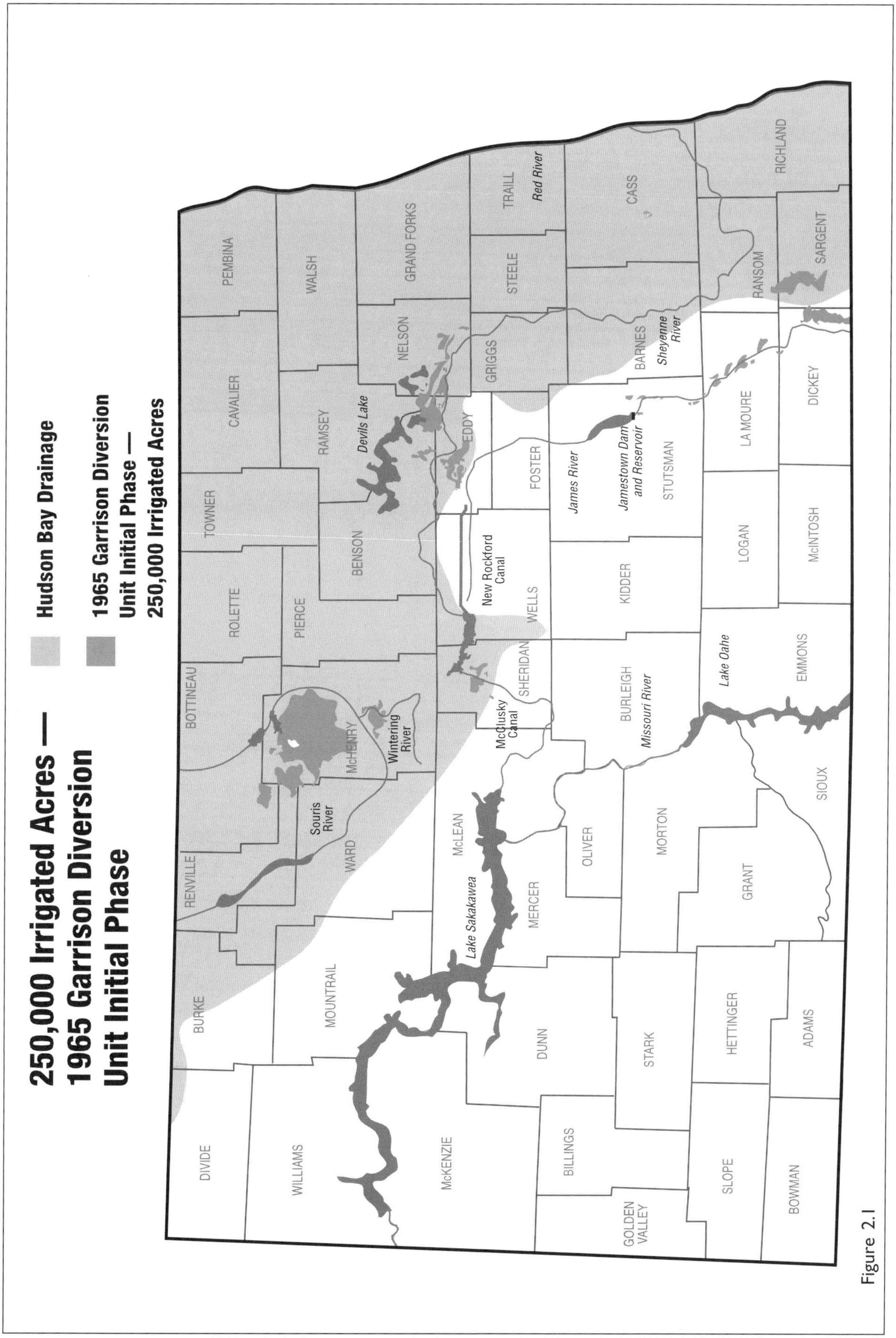
250,000 Irrigated Acres —
1965 Garrison Diversion
Unit Initial Phase
Hudson Bay Drainage
1965 Garrison Diversion
Unit Initial Phase —
250,000 Irrigated Acres
DIVIDE
BURKE
RENVILLE
BOTTINEAU
ROLETTE
TOWNER
CAVALIER
PEMBINA
WALSH
GRAND FORKS
TRAILL
Red River
CASS
RICHLAND
SARGENT
RANSOM
STEELE
NELSON
RAMSEY
Devils Lake
GRIGGS
BARNES
Sheyenne River
EDDY
FOSTER
James River
Jamestown Dam and Reservoir
STUTSMAN
LA MOURE
DICKEY
McINTOSH
LOGAN
KIDDER
WELLS
New Rockford Canal
BENSON
PIERCE
McHENRY
Wintering River
Souris River
WARD
MOUNTRAIL
WILLIAMS
McKENZIE
McLEAN
Lake Sakakawea
McClusky Canal
SHERIDAN
BURLEIGH
Missouri River
Lake Oahe
EMMONS
SIOUX
MORTON
OLIVER
MERCER
GRANT
DUNN
STARK
HETTINGER
ADAMS
BILLINGS
GOLDEN VALLEY
SLOPE
BOWMAN

Figure 2.1

Figure 2.2

Missouri River Diversion Milestones

early 1800s	American Fur Company transports cargoes.
1838	U.S. Army Corps of Engineers first involved in improvement.
1881	Missouri River Improvement Convention.
1894	Carey Act - federal land offered to settlers if irrigated.
1902	Reclamation Act - financing irrigation works.
1904	Mandan bluffs chosen as site for dam.
1906	North Dakota pumping project authorized.
1909	Boundary Waters Treaty signed.
1922	North Dakota Irrigation Congress meets in Bismarck, June 7-8.
1924	Missouri River Diversion Association created in Devils Lake.
1927	Elwyn F. Chandler plan for diversion, University of North Dakota.
1928	Flood Control Act - Corps surveys Mississippi tributaries, chooses Fort Peck as dam site.
1934	North Dakota Legislature petitions Federal government for diversion. Corps produces study of Missouri River problems.
1935	Water Conservation District Law passed.
1936	Flood Control Act - prevention of floods becomes national concern. Great Plains Committee established - initial surveys and recommendations.
1937	North Dakota State Water Conservation Commission created.
1940	Pick (Corps of Engineers) and Sloan (Bureau of Reclamation) Plans completed. Pick plan contains Garrison Reservoir.
1941	Missouri Valley Regional Planning Commission created.
1944	Senator James Murray of Montana sponsors Missouri Valley Authority legislation. Corps and Bureau compromise—Pick-Sloan Plan incorporated into Flood Control Act. Plan includes Garrison Reservoir.
1945	Missouri Basin Inter-Agency Committee created.
1947	Construction on Garrison Dam begins.
1948	Soil surveys determine Crosby-Mohall Unit unsuitable for irrigation.
1953	Garrison Dam closure ceremony. Bureau releases Garrison Diversion Plan.
1954	Wildlife groups critical of Garrison Diversion. Interim report on Garrison Diversion completed and approved.
1957	Bureau submits Garrison Diversion Unit plan to Interior. Congressman Otto Krueger introduces House Bill 7068 to authorize Garrison Diversion.
1959	Bureau of the Budget against Garrison Diversion because of cost.
1961	Garrison Diversion Unit modified to 250,000 acres. Authorization fails in Congress.
1965	Garrison Diversion authorized at 250,000 acres.
1968	Snake Creek Pumping Plant construction begins.
1970	National Environmental Policy Act requires environmental impact statement. Canadian government expresses concern in aide-memoirs over return flows (biota).
1972	Committee to Save North Dakota files suit to halt construction.
1973	Canadian government requests moratorium.
1974	Final EIS filed. Environmental groups challenge Garrison Diversion Unit.
1975	Garrison Diversion Unit referred to International Joint Commission.
1976	National Audubon Society files suit against Interior alleging EIS violated NEPA.
1979	Bureau of Reclamation begins biota transfer studies.
1981	Attempt to block funding in Senate fails. New Rockford canal criticized.
1982	James River Flood Control Association files suit, temporary restraining order.
1983	Restraining order removed.
1984	Focus of Garrison Diversion Unit shifts from irrigation to municipal water delivery. Lonetree Reservoir deferred.
1986	Garrison Diversion Reformulation Act reauthorizes 130,000 acres irrigation project.
1988	Interbasin Biota Transfer Studies Project begins.
1993	Garrison Diversion Conservancy District manager proposes revisioned plan.
1997	Work continues on proposed amendments to the Garrison Diversion Reformulation Act of 1986.

their cargoes eastward. As early as 1838, the U.S. Army Corps of Engineers became involved in improving transportation on the river by removing snags in the river (Hart 1957).

A general optimism of growth in the Missouri basin in the 1880s produced a demand by its residents to revise a failing river commerce due to competition from the railroads. In 1881, three hundred delegates attended the Missouri River Improvement Convention at St. Joseph, Missouri, to discuss stimulating river development. They sought the means to stabilize the silt-laden stream by narrowing and deepening the channel. In 1884, in response to the convention, Congress placed all of the functions of the Corps of Engineers on the Missouri River under the guidance of the Missouri River Commission (Hart 1957).

The Commission was to administer and direct improvements on the Missouri River and to estimate and maintain channel depth on the river for the purpose of commerce and navigation. From the beginning, however, the commission was plagued by political and economic problems. Individual, corporate, and municipal interests desired protection of their property along the banks of the river. By 1897, the Commission became hindered by reduced Congressional appropriations and widely separated locales, all seeking improvements. In 1902, the Commission dissolved, and works along the river were left unfinished. The decline in interest in the Missouri River Commission coincided with years of low water in the river, agricultural depression, and population loss (Hart 1957).

Using the waters of the Missouri River for irrigation became a possibility in the late nineteenth century when a new wave of settlement began. A series of legislative acts served as the stimulus for this new wave. The Homestead Act of 1862 permitted any adult citizen, or adult alien seeking citizenship, to select 160 acres in the public domain. *The Desert Land Act of 1877* gave title to 640 acres of arid land, provided the settlers put water on it. The Desert Land Act was the first major effort by the federal government to encourage irrigation (Whittlesey 1986).

By the 1890s, the Dakotas enjoyed an influx of new agricultural settlers, motivated by the belief they would prosper in light of a mounting scarcity in the nation's food supply and rising farm prices. Irrigation came to be seen as a cure-all for the climatic problems of the area. In 1891, Congress appropriated money for an irrigation survey, and irrigation enthusiasts sought direct help from the federal government to build projects. The Carey Act of 1894 offered federal lands to settlers on the condition they irrigate and settle on it. Few farms, however, went "under the ditch." The drought of the 1890s peaked in 1896; however, irrigation could continue only if the federal government offered more direct help than land subsidies (Hart 1957).

Help came in the form of the Reclamation Act of 1902, which has since served as the foundation for reclamation in the United States. The Act established a reclamation fund for the purpose of financing irrigation works from the sale and disposal of public lands (Whittlesey 1986). By 1903, North Dakota was looking seriously at irrigation. In the summer of that year, F. E. Weymouth was sent by the newly created Reclamation Service to find suitable projects in North Dakota. Weymouth found few suitable areas in the state; but because North Dakota had contributed significantly to the reclamation fund, a project had to be found (Hart 1957).

The state re-established the office of State Engineer and appropriated funds to investigate the practicality of irrigation. The railroads, Minneapolis and St. Paul businessmen, and local bankers offered support to organize a State Department of Irrigation and Engineering to conduct preliminary surveys for irrigation development (Kelly 1989). By August 1904, the Mandan Bluffs between Bismarck and the mouth of the Little Missouri River had been chosen as a site for a huge dam. The two-hundred-foot high dam would create a lake two miles wide and two-hundred-fifty miles long. The project would provide flood control, water for irrigation and navigation improvement, and would create electrical power (Harrison 1904).

Irrigation suffered a temporary setback in late 1904, when Chief Engineer F. D. Newell, of the Reclamation Service, concluded that no considerable irrigation project was practical in North Dakota, except for about sixty or seventy thousand acres near Buford (Kelly 1989). In an attempt to attract more settlers to the region, the railroads, realizing the possible profits from increased tonnage and passengers, began to show more interest in irrigation.

The State Fair in Mandan on September 27, 1904, offered the railroads the opportunity to exhibit irrigation and irrigated products. The North Dakota Irrigation Congress invited L. W. Hill of the Great Northern Railway, President Howard Elliot of the Northern Pacific Railway, and George Maxwell of the National Irrigation Association to praise the benefits that irrigation would bring to the area (Kelly 1989). However, North Dakotans appeared uninterested or doubtful of irrigation. The September meeting of the North Dakota Irrigation Congress was poorly attended, and the state had only two or three small irrigation projects operating at the time (Kelly 1989). The doubts and lack of interest did not prevent the first federal government irrigation project in North Dakota from being authorized in January 1906.

The North Dakota Pumping Project consisted of three projects: the Williston Project, the Nesson Project, and the Buford-Trenton Project. Work began first on the Buford-Trenton Project, but early problems developed on the other projects. The Nesson Project operators had signed a contract setting the building charges, but the contract did not meet government requirements. Many of the original signers refused to sign a second contract because the crops did not need irrigation water, due in part to generous rainfall.

By 1909, the problem of construction costs became serious (Kelly 1989). According to reclamation law, construction charges were to be determined with the view of returning the estimated cost of construction to the reclamation fund. The contracts signed between the government and the landowners were based on actual cost instead of estimated cost. Farmers pledged their property to guarantee fulfillment of the contract, resulting in the confiscation of land and homes by the federal government in the event of project bankruptcy.

By January 1910, the Reclamation Service became weary of trying to settle the problems of repaying construction charges with the Williston project and prepared to withdraw from the project. Farmers on the Williston project also became discouraged. The works proved to be much more expensive than had been estimated, and the charges being assessed by the Reclamation Service were so excessive that irrigators could not prepare their lands properly for the water. Other problems included farmers' lack of experience with irrigation and generous rainfall. As a result, many farmers refused to accept any water, and the early enthusiasm for irrigation died out quickly (Kelly 1989).

The Reclamation Service attempted to solve the problem by extending the payment deadline and offered financial help to operators to make the water payment, but the interest was twelve percent. Adequate rainfall in 1912 meant that little or no irrigation water was required for the projects. The project struggled until 1915, when it closed because farmers were unable to pay the annual water charges. An irrigation district was formed again in 1918 and requested the Bureau of Reclamation to resume its activities. The project operated through 1925, when it was sold to private parties (Kelly 1989).

North Dakota's first attempt at irrigation was disastrous. Some of the reasons listed for the demise of the project by the early operators were

- the failure of wheat from rust due to excessive moisture,
- the high charges for water,
- the lack of a market for vegetables, and
- the cost of preparing the land for irrigation.

Perhaps the greatest factor in the failure of the project was that the people were not "irrigation-minded" (Kelly 1989).

The development years (1915-1945)

From 1914 to 1919, the demand for food to supply the nation, and its wartime allies, placed a renewed emphasis on grain production. North Dakota's role was to supply the wheat; and although the price of wheat rose, the crops were poor due to drought and rust. Land values increased, however, and many farmers borrowed heavily to buy land to take advantage of high prices. When land values and crop prices began to decline in the 1920s, many farmers lost their farms, and many banks failed. The problem of scant rainfall also persisted (Robinson 1966). The first reaction to the drought was a revival of interest in irrigation.

During the period 1917-21, the state revised or established new codes regulating the formation and operation of irrigation districts. North Dakota sought the help of Montana and South Dakota in 1922 to revive interest in using Missouri River water for irrigation. Proponents of irrigation wanted the North Dakota Irrigation Association, which had dissolved after the failure of the North Dakota Pumping Project, revived. George H. McMahon, State Engineer, called for a general irrigation congress to consider financing some small individual irrigation projects (Kelly 1989).

A bill approved by Congress in May 1922, made $20 million available for irrigation in North Dakota. Plans were hastily made for an irrigation congress to be held June 7-8 in Bismarck (Kelly 1989). The meeting was held, but the results were disappointing. Only twelve counties from North Dakota were represented, and there were no representatives from Montana or South Dakota. The congress split into two factions: those who supported small irrigation projects, and those who supported large irrigation projects. The congress developed a plan to have the state engineers investigate the possibility of small tracts, while having the federal government explore the possibility of large projects (Kelly 1989).

Despite some enthusiasm for irrigation among boosters, North Dakota agriculturalists remained generally uninterested. This can be attributed to a number of factors:

- Dry farming usually returned decent crops.
- Those who did irrigate complained of excessive pumping costs for water.
- The distance from markets discouraged truck farming.
- Irrigated grain was generally inferior in quality.
- Yields were not great enough to offset costs of irrigation (Hargreaves 1957).

Despite problems with irrigation and the apparent lack of interest in it, the diversion of water from the Missouri River for purposes other than irrigation remained an important topic across the state. In 1924, Sivert W. Thompson, Secretary of the Devils Lake, North Dakota, Chamber of Commerce and an advocate of Missouri River diversion, and the people of Devils Lake formed the Missouri River Diversion Association. Thompson and the Association wanted water diverted to Devils Lake to raise its level, which had been falling gradually, and to sweeten the water, which had grown increasingly alkaline as the level had dropped (Robinson 1966).

Although the idea of diversion of Missouri River water dates back to the State Constitutional Convention in 1889, no documented plans existed until 1927. In that year, Elwyn F. Chandler of the University of North Dakota, Grand Forks, devised a plan to divert the water. Chandler wanted a concrete-lined tunnel to divert the water for irrigation purposes and to restore Devils Lake. Although he could not estimate the cost, he warned the state not to be too slow to act (Kelly 1989). Chandler's diversion plan would transfer Missouri River water across the Continental Divide into the Hudson Bay basin. At the time, there was no concern about possible environmental damage or biota transfer that might result from the diversion of water from one drainage basin to another.

Mississippi River floods in 1927 caused extensive damage and created a national demand for Congress to investigate all techniques for river control and all uses of river water. The Flood Control Act of 1928 directed the Chief of Engineers to survey the major tributaries of the Mississippi. The Corps of Engineers took a scientific approach to the problem by preparing hydrographs and plotting the flow of the river. Two important developments happened: the creation of reservoirs became an accepted civil engineering practice and the Missouri River began to exhibit extremely low flows. Reservoir storage now became a necessity for maintaining channel depth for navigation. Fort Peck in Montana was chosen as a site for a great dam (Hart 1957).

The Corps of Engineers was undecided about the capacity of the dam. Should it be built for 6,400,000 acre-feet of water for a six-foot channel, 17,000,000 acre-feet for a nine-foot channel, or abandon the project altogether? The Chief of Engineers recommended Fort Peck be built for the maximum capacity. The dam was authorized by Franklin D. Roosevelt in the National Industrial Recovery Act in 1933. Congress did not act on the dam until 1938, when Fort Peck was inserted into the Rivers and Harbors Bill for 1940. Roosevelt then vetoed the bill because the non-military activities of the War Department had to give way to military preparedness (Hart 1957).

The years of the 1930s were especially important to North Dakotans regarding the use of Missouri River water. The Great Drought devastated North Dakota. Nine years, between 1929 and 1939, had less than average rainfall. Hot winds and searing heat destroyed crops, and the lack of pasture grasses devastated livestock. Depression prices made the drought years even worse. Low income caused a decline in land values, an increase in foreclosures and tenancy, and growing corporate and government ownership of land. Large numbers of people went on relief, and a rural exodus occurred (Robinson 1966). North Dakotans became more conscious than ever of a shortage of water, and their attention focused on the Missouri River as a possible solution.

After the drought of 1934, the North Dakota Legislative Assembly petitioned the federal government to create a diversion using Missouri River water. The state proposed a dam between Mannhaven and Garrison to restore groundwater tables, supply surface waters, and control flooding (Kelly 1989). The state also revised its irrigation laws and, in 1935, passed the Water Conservation District Law, which authorized a conservation district to build dams, reservoirs, and regulate the use of water (Robinson 1966).

Severe flooding in 1935 brought passage of the Flood Control Act of 1936. The Act recommended the prevention of floods as a national concern. It required local interests, who benefited from construction works to pay for lands, easements, and damages involved, and authorized the Department of Agriculture to engage in soil erosion prevention (Hart 1957). Irrigation projects appeared in every county where possible in North Dakota. More than 2,000 dams were installed in the western counties of the state in the fall of 1937 (North Dakota Cooperative Extension Service 1937).

Also in 1937, the State Legislature created, by emergency action, the North Dakota State Water Conservation Commission to develop, restore, and stabilize waters for domestic and agricultural use, municipal purposes, irrigation, recreational needs, and wildlife conservation through the construction of dams, reservoirs, and diversion canals. The Commission was to finance both public and private works to accomplish these goals (Robinson 1966). By 1939, the State Water Commission subsidized engineering costs and supplied pumps and pipes to farmers who prepared their land for irrigation.

Politically, the time was right for a comprehensive Missouri River Valley development program. The droughts of the 1930s demonstrated the need for better water management. In 1938, Congress authorized nine reservoirs on Missouri tributaries. In 1941, two more were authorized, and levees were to be built along the river from Sioux City to Kansas City. Appropriations had not been requested when World War II began, which slowed the development of water resource programs (Hart 1957).

Ample rainfall and good farm produce prices brought prosperity to North Dakota during and after World War II. Farm production increased, yet many farmers gave up farming. Farm population and the number of farms decreased while the size of farms increased, primarily because of mechanization. The rural exodus continued to worry state planners. Returning servicemen could not expect to find much work in agriculture, but Missouri River development held some promise. The North Dakota Reclamation Association, the Greater North Dakota Association, the North Dakota Farm Bureau, and chambers of commerce all supported irrigation to aid economic diversification (Kelly 1989). The reason for the renewed interest was not just to stop the rural exodus. The possibility of the federal government spending huge appropriations in North Dakota had become a reality.

Following the disastrous drought of 1934, the Army produced its first monumental study of Missouri River problems (Ridgeway 1955). In 1936, President Roosevelt established the Great Plains Committee to make initial surveys and recommendations regarding the Missouri River. Their preliminary report appeared in 1938, and the comprehensive report in 1940. The Committee recommended expenditures for irrigation projects, with detailed planning to be undertaken by various federal agencies with "interests" in the area. By 1940, the Army Corps of Engineers had completed a study of the feasibility of controlling the Missouri River to minimize flood damage to downstream states. The Bureau of Reclamation also worked on a plan to divert waters for irrigation purposes (Ridgeway 1955).

In May 1941, the Missouri Valley Regional Planning Commission was established to formulate policies according to the National Resources Planning Board recommendations. The Commission was composed of one official delegate each from Iowa, Kansas, Missouri, Minnesota, Montana,

Nebraska, North Dakota, and South Dakota. Also included was one representative each from the Department of War, the Department of Agriculture, and the Department of the Interior. The Commission's first crisis involved the problem of basin development priorities. The Planning Board did not recommend large-scale irrigation projects, because they did not yield substantial benefits in wet years and fell into disuse (Ridgeway 1955).

The Army's plan, also known as the Pick Plan, after its author Colonel Lewis A. Pick, called for a series of six giant dams on the Missouri to control flooding and improve navigation. Also included were twelve additional multi-purpose reservoirs between Yankton, South Dakota, and Garrison, North Dakota. A reservoir at Garrison would allow diversion of water across the basin divide to the Devils Lake area and the headwaters of the James River (Ridgeway 1955).

The Bureau of Reclamation plan, known as the Sloan Plan, after its author Warren Glen Sloan, accepted the Army's plan for flood control and navigation proposals on the lower river but modified the upper river plan. The Bureau wanted to use the reservoirs for purposes other than flood control and navigation, namely hydroelectric power and irrigation. A proposed Missouri-Souris Diversion Project would divert water for irrigation by canal from the Fort Peck Reservoir in Montana. Sloan also wanted to recharge Devils Lake and divert the runoff into the Red River for domestic water supplies. The Sloan Plan did not include a dam and reservoir at Garrison (Ridgeway 1955).

Beginning in 1940, Pick and Sloan traveled extensively throughout the Missouri Basin to promote their rival plans. Although both plans were well received by the basin states, they came increasingly under attack by a number of agencies. The Bureau of Reclamation criticized the Army for not being comprehensive enough, since its aim was only flood control and navigation improvement. The Army criticized the Bureau's plan because it questioned the Missouri-Souris Diversion and was against large-scale irrigation. The Bureau of the Budget criticized both plans because of the tremendous cost, an estimated $1.2 billion (Ridgeway 1955).

The rival agencies disagreed sharply over the location of a reservoir at Garrison. Sloan opposed a dam and reservoir there because it would only slightly increase water flow for navigation, the cost was too great, and it was not needed for any other purposes. It would increase storage, but all the advantage gained would be lost by evaporation. The Department of Interior opposed Garrison because most of the Fort Berthold Indian Reservation lands would be flooded (Ridgeway 1955).

The basin states became more involved in July 1942, when the Governors of Iowa, Kansas, Missouri, Montana, Nebraska, North Dakota, South Dakota, and Wyoming met in Montana to form the Eight States Committee, later renamed the Missouri River States Committee. Their goal was valley-wide development, and they openly endorsed both plans (Ridgeway 1955). Spring floods in 1943, and heavy rain in June, brought the problems relating to the Missouri River to the attention of Congress.

Congressional hearings on the Pick Plan began in the spring of 1943. Spokesmen for the upper basin states, reclamationists, and irrigationists viewed the plan as a threat to Great Plains development and irrigation. They did not want the Pick Plan rushed through Congress until the Sloan Plan could be heard as well. As the hearings continued, the conflicts between the two plans became apparent. The navigation-irrigation conflict, state's water rights, the need for additional irrigated land, and whether there was a large enough population in the basin to warrant additional production of hydroelectric power were openly discussed (Ridgeway 1955). The Missouri-Souris Diversion raised the question of return flows and biota crossing international boundaries. At the request of the Bureau of Reclamation, amendments were added to the bills to address the conflicts. The O'Mahoney-Millikin amendments provided that Congress recognize states' rights to improve watersheds within their boundaries. The upper basin states were to have priority water use for domestic, municipal, irrigation, and industrial purposes over those of the downstream states for navigation (Ridgeway 1955).

Intensive lobbying began in 1943, and matters became even more confusing when Senator James Murray of Montana introduced legislation in 1944, to establish a Missouri Valley Authority (MVA). Modeled closely after the Tennessee Valley Authority, Murray's bill aimed to provide a central authority for unified water control and resource development. The National Farmers' Union and the North

Dakota Farmers' Union both supported an MVA. President Roosevelt also supported a central authority for development. Public power groups opposed it because they believed the sale of hydroelectric power was restricted by the amount of water available for power production. Others saw MVA as another move to socialize America (Ridgeway 1955).

The congressional elections in 1944 stalled the MVA bill, and its final defeat was probably certain when the Bureau and the Army put their rivalry aside and reached a compromise between their respective plans. The two plans were combined to form the Pick-Sloan Plan, also called the Missouri Compromise, or as the North Dakota Farmers' Union called it "a shameful, loveless shotgun wedding." The new plan gave the Army the responsibility for determining the location of the main stem reservoirs. The Bureau was responsible for the location and capacities of the minor reservoirs and the probable extent of irrigation. Both agencies recognized the need for full development of hydroelectric power. The Pick-Sloan plan also included the Garrison Reservoir, even though Sloan believed there was not enough water to satisfy all the needs of the basin (Ridgeway 1955).

To serve as an informational, administrative agency, the Missouri Basin Inter-Agency Committee was created on March 29, 1945. It was organized along the lines of the Missouri River States Committee, but with no legal authority or appropriation. Membership was voluntary and was composed of various national agencies working on Missouri River problems, as well as basin state representatives. The agency proved ineffective almost from the beginning and served chiefly as a forum for the basin states to discuss problems and hear reports on the progress of basin development (Ridgeway 1955).

The problem years (1946 to date)

W. G. Sloan's plan for irrigation, through the Missouri-Souris Diversion Unit, would bring enough water from the Fort Peck Dam in northeastern Montana to irrigate approximately one million acres in the Crosby-Mohall Unit. Located in portions of Divide, Burke, Ward, Renville, Bottineau, and McHenry counties of northwestern North Dakota, the Crosby-Mohall Unit was the largest unit of the project that was to receive irrigation water. Water would be diverted into the Missouri Canal to fill the Medicine Lake Reservoir. Water would then be lifted by a pumping plant near Grenora into the Souris Canal. Sloan hoped to irrigate an additional 55,000 acres in the New Rockford Unit, located in Eddy County; 22,000 acres in the Jamestown Unit, located in Stutsman County; and 30,000 acres in the Oakes Unit, located in Dickey and Sargent counties. Return flows would restore the levels of Devils Lake and Stump Lake (Missouri Basin Inter-Agency Committee 1952). Water diverted into the Souris River would also cross the border into Canada, violating the Boundary Waters Treaty of 1909. The Bureau of Reclamation continued topographical, geological, and land classification surveys through 1946 and into 1947. Small irrigation farms were established in the Crosby-Mohall area to obtain information on the operations of the project and to determine the benefits of irrigation. Construction costs had also nearly doubled since 1940 (Richardson 1973).

Political developments after 1947 began to pose a threat, not only to the Missouri River Basin Development Program, but to all land and resource projects across the country. Interior Secretary Julius Krug, under President Harry S. Truman, was faced with the problem of preserving resources and administering water and power programs. The public's postwar desire for unlimited growth was clashing with a movement to cut funding for federal land and resource projects. The result was a "no new starts" policy that cut funding for most projects already begun and no funding for projects not yet begun (Richardson 1973).

Bureau of Reclamation investigations continued through 1949, focusing on the Bowbells Block, a tract of land in the vicinity of Bowbells in Burke County. The Bureau of Reclamation believed their research would provide answers to almost every agricultural problem to be encountered on North Dakota soils. The Bureau was anxious to get the Fort Peck diversion dam started, but soil surveys conducted in 1948 indicated that most of the original area targeted for irrigation by diversion from the Fort Peck Dam was unsuitable because of drainage problems caused by the density of the glacial subsoil.

On March 22, 1950, Kenneth Vernon, Director of Region 6 of the Bureau of Reclamation, made the announcement that some 500,000 acres west of the Des Lacs River were disqualified. The Bureau shifted its attention eastward, where another 2.5 million acres had to be investigated to find enough suitable land to justify the Missouri-Souris Diversion Project (Kelly 1989).

On March 28, 1950, the Bureau revised its plans concerning the Missouri-Souris Project. The plan was changed because of a temporary delay in the construction of the Missouri-Souris Diversion Dam near Fort Peck and the results of the negative unacceptable drainage characteristics of the soils found in the Bowbells Block. They began considering either a new structure or the modification of an existing structure that, when completed, could also be used to produce and sell hydroelectric power. If found feasible from an engineering standpoint, power revenues would be used to assist in paying the cost of diversion in future years (Kelly 1989). Modifications to the hydroelectric component of the Garrison Dam, already under construction, was the option selected. In August 1950, the Bureau halted investigative work on the one-million acre Crosby-Mohall Unit because of the recent findings. In October, the Bureau reported that 1.5 million dollars had been allotted for an investigation east of the Bowbells Block to identify four to five million acres for suitable irrigation projects. A new plan was expected to be finished by September 1953 (Kelly 1989).

As of December 1950, the Missouri Basin Development Project was moving very slowly. Many opponents began questioning such a large expenditure, given the fact that very few benefits were being realized. W. G. Sloan believed that part of the problem was the fact that policy seemed to require each project to be justified on an individual basis. This caused a stretching of the facts to prove overall project feasibility. Another problem was the negotiation of irrigation repayment contracts. One area of particular concern was the Heart River Project, where farmers had begun legal action to be excluded from the project (Kelly 1989).

In the election of 1952, Dwight D. Eisenhower led a huge Republican victory nationwide. Eisenhower promised to alter twenty years of domination by the federal government of water and power development projects. Eisenhower believed the federal government should undertake only those projects which the states could not do themselves. He did not oppose comprehensive river basin development, but wanted the states and local interests to accomplish it themselves. He chose to achieve this goal, not by formulating a fully planned program, but by an initial phase of routine decisions on pending projects. Appropriations for the Bureau of Reclamation, already frozen by Truman, were not increased for the next fiscal year. All new starts were eliminated for 1954, and current construction slowed or stopped wherever it was not far advanced (Richardson 1973).

Eisenhower announced that the budget for 1953 would allot half of the total requested for water resource projects and multiple purpose dams because of their importance in national emergencies (Richardson 1973). The giant dam at Garrison, started in 1947, was nearing completion and was safe from having its appropriations cut. There was no plan yet for diverting water from Garrison Reservoir. Basin development as a whole was in jeopardy, but North Dakota had a couple of officials in key government positions. Fred Aandahl, former Governor, creator of the State Water Commission and former member of the Missouri Basin Inter-Agency Committee, was now Assistant Secretary of the Interior and in charge of the Interior's water and power projects. Senator Milton R. Young, a long-time political friend of Aandahl's, became chairman of the Rivers and Harbors Appropriation Committee. Both were strong supporters of Missouri River development.

In April 1953, the Bureau of Reclamation released a report advocating diversion of water from Garrison Reservoir. Studies had been underway since 1940 to produce a plan. According to Bureau estimates, it was cheaper to divert water from Garrison than from Fort Peck. The area being considered for irrigation was at least a million acres, and could be served from a single water supply system diverted from the Snake Creek Arm of the Garrison Reservoir (Kelly 1989).

Garrison Diversion also offered the prospect for additional water in most of the rivers in the eastern part of the state, a chance to raise and regulate Devils Lake, an improvement in marsh and pothole habitat for migratory waterfowl, and an increase in surface water supplies for municipal and industrial use. Pollution abatement from the diversion would help dilute stream pollution by increasing flows. The Bureau believed that the opportunity for community growth and development would go far beyond their ability to measure in terms of dollars (Kelly 1989).

By 1954, Missouri River development was being criticized because the Corps of Engineers and the Bureau of Reclamation had failed to mitigate wildlife habitat flooded by the dams across the state (Kelly 1989). Farmers from across the state had generally been uninterested or indifferent to irrigation, but now wildlife groups became more vocal and involved because of the possible damage to wildlife habitat.

By October 1954, the Interim Report on Garrison Diversion had been completed and approved. The cost was estimated at $435,000,000. This figure did not include any allocations for the cost of the main stem reservoirs. Eighty percent of the cost would come from the surplus revenues generated by the sale of hydroelectric power. The Bureau of the Budget set the benefit to cost ratio at 0.7 benefits to 1.0 for costs (Kelly 1989). Hopes were high across the state for the benefits to be realized from irrigation. Aandahl predicted that small family farms would be re-established, new jobs would bring 93,000 new people to the state, and state income would increase substantially (Kelly 1989). The North Dakota State Legislature created the Garrison Diversion Conservancy District (GDCD) in 1955 as the overall organization to be concerned with the Garrison Diversion Unit. The 25-county political subdivision of North Dakota encompassed 45 percent of the state's area, 60 percent of its population, and 60 percent of its taxable valuation. The GDCD brought all the various interests in the project into one responsible organization and provided a means whereby North Dakota's basic industry—agriculture—would be benefited through the development of irrigation of carefully selected lands scattered throughout the central and eastern sections of the state. Only 45,000 acres were under irrigation by 1956, with 30,000 acres located in McKenzie and Williams counties (North Dakota Extension Service 1956).

In January 1957, the Bureau submitted the Garrison Diversion Unit plan to the Department of the Interior. The water would be lifted by huge pumps from the Garrison Reservoir to the Snake Creek Reservoir. The McClusky Canal would carry the water 73 miles to the Lonetree Reservoir, at the headwaters of the Sheyenne River. Canals from Lonetree Reservoir would fill Devils Lake, and transport irrigation water to the proposed areas. The project would be built over a period of 60 years and contain 6,773 miles of canals, eight reservoirs, water supplies for 41 towns, 656 pumping stations, and 9,300 miles of drain, at an estimated cost of $529,000,000. Secretary of the Interior, Fred Seaton, approved the plan in June (Robinson 1966).

In October 1957, North Dakota Congressman Otto Krueger introduced House Bill 7068, which amended the Flood Control Act of 1944 to authorize Garrison Diversion. A House subcommittee, chaired by Rep. Wayne Aspinall of Colorado, met in Devils Lake to hear testimony from interested parties. The testimony was mostly in full support of Garrison Diversion, but Aspinall cautioned the supporters that irrigation would take many years to come and the cost was growing exceedingly high. Other congressmen at the hearing were concerned about the water from one watershed emptying into another, such as the Souris and Red Rivers emptying into Canada. The huge cost of the project was also delaying its approval in Congress (Kelly 1989).

Secretary Seaton did not send it to the Bureau of the Budget until October,1959. Acting Director of the Bureau of the Budget, Elmer B. Staats, recommended it not be passed because of the enormous cost and the dependence on secondary benefits. It did not reach Congress again until 1960 (Robinson 1966). In June 1961, disapproval in Congress over the cost resulted in a modification of the project. Irrigation was reduced to 250,000 acres, water was provided for municipal and industrial water supplies for fifteen towns, twenty-four areas for fish and wildlife conservation, and recreational development at seven major water impoundments (Kelly 1989). Hearings held before the Subcommittee on Irrigation and Reclamation, of the Senate Committee on Interior and Insular Affairs, failed to reauthorize the project. It failed again in 1963 and 1964 because of the cost (Josephy 1975).

In 1965, the 250,000-acre initial stage of the Garrison Diversion Unit was reauthorized, and in 1966, the State of North Dakota and the United

States signed a master contract to build the unit. Actual construction on the Snake Creek Pumping Plant began in 1968 (Kelly 1989). On January 1, 1970, the National Environmental Policy Act (NEPA) took effect. All agencies of the federal government were required to include, in every report on a project, an Environmental Impact Statement (EIS) detailing its effect on the quality of human life, especially the environment. The Garrison Diversion Project was now threatened because of its possible effect on the environment.

Canadian concerns over the Garrison Diversion Unit were expressed in an aide-memoirs in 1970. Manitoba was concerned that leaching of irrigated soils would degrade the water quality of shared watersheds and that return flows would increase flooding. The possible introduction of biota, such as foreign fish and eggs, parasites and diseases, would impact existing aquatic systems and commercial and recreational fishing (International Joint Commission 1977). Various groups across the state began seeking a one-year moratorium on the project because the final EIS had not been filed. In December 1972, the Committee to Save North Dakota, Inc. filed suit in U.S. District Court seeking a halt to further land acquisition by the Bureau of Reclamation and a halt to construction (Kelly 1989). In October, 1973, the Canadian Government, with the support of the Manitoba Naturalists Society, Canadian House of Commons, and the Canadian Department of External Affairs, also requested a moratorium on the project (Kelly 1989). The project became severely criticized by many state and national organizations after the Bureau of Reclamation filed the final EIS in January 1974. Wildlife societies, ecology groups, and environmental groups all challenged the project (Kelly 1989).

In 1975, the Canadian and United States governments referred the matter to the International Joint Commission to report on the impact of Garrison Diversion and make recommendations to ensure that Article IV of the Boundary Waters Treaty of 1909 was honored. The Commission established the International Garrison Diversion Study Board to undertake technical investigations to assess the impact of the project. In November 1975, the International Joint Commission held public hearings concerning the impact of the project on Canada. Most of the testimony was against further construction of the project. Based on testimony from this, as well as seven other hearings and the Study Board Report, the Commission concluded that construction and operation of the project would have an adverse effect on Manitoba, causing injury to health and property as a result of adverse impacts on the water quality and biological resources of the province. Modifications to eliminate direct connections between the Missouri River and the Hudson Bay drainage basin would reduce some of the adverse effects, but the Commission recommended that the portion affecting water flowing into Canada, Lonetree Reservoir, not be built at that time (International Joint Commission 1976).

In May 1976, the National Audubon Society filed suit against the Department of Interior, alleging that the EIS on Garrison violated the NEPA. An out-of-court agreement promising a new EIS was reached in May 1977. In reaction to the Commission's recommendation, President Jimmy Carter included Garrison Diversion in his list of unsupportable water projects and recommended no further funding; however, the Garrison Diversion Project was later taken off the list (Kelly 1989). The Interior Department issued a new EIS in January 1978, which presented six alternative plans for Garrison Diversion. The plans ranged from the original 250,000-acre project, authorized in 1965, to a plan which required termination of construction. Lonetree Reservoir was still included in the plans at the insistence of North Dakota's political leaders (Kelly 1989).

In August 1978, the U.S. Senate rejected a request by the Carter Administration to delete $18.7 million from the project. The Interior Department quickly resumed work on the project, resulting in renewed litigation by the National Audubon Society. On May 6, 1981, the U.S. District Court ruled that the project could continue. The Audubon Society appealed in May 1982, but the U.S. Court of Appeals upheld the decision (Kelly 1989).

An attempt to block further funding in the Senate failed in 1981, but opponents continued to attack individual projects, such as the New Rockford Canal. The Canal was to transport water from the already controversial Lonetree Reservoir to the James River Canal, where the water would irrigate the Oakes Unit in southeastern North Dakota.

Landowners along the proposed route of the canal were against the canal because they believed the Bureau of Reclamation should use the James River as the conveyance (Kelly 1989).

Negotiations continued until May 1982, when the James River Flood Control Association in South Dakota filed suit in U.S. District Court. The Association, with the help of the National Audubon Society, contended that the project would have detrimental effects on the James River. U.S. District Judge Donald Porter agreed and issued a temporary restraining order on May 18, 1982, preventing any further land acquisition. But in January 1983, Judge Porter removed the restraining order, citing the EIS as adequate (Kelly 1989).

In 1984, the Garrison Diversion Commission, a federal commission appointed by the Secretary of the Interior, recommended the focus of the project be shifted from irrigation to municipal water delivery, and that the controversial aspects of the project be deleted, most notably Lonetree Reservoir. The proposed Sykeston Canal would provide the water supply link between the McClusky and New Rockford Canals. The Lonetree area would be managed for wildlife by the State of North Dakota in the interim. All irrigation would be restricted to lands that drain, or could be drained, into the Missouri River basin (Bureau of Reclamation 1986). The proposal would irrigate lands in the Hudson Bay drainage.

The deferral of Lonetree brought new hope for the project; and in 1986, Congressman Byron Dorgan sponsored H.R. 1116 to reauthorize the project. The amended version reduced irrigation acreage from 250,000 acres to 130,000 acres, with emphasis on municipal, industrial, and rural domestic water supplies. The 1986 Act was supported by the conservation groups. One of the specific features of the reauthorization was the Sykeston Canal, which replaced Lonetree Reservoir as a conveyance for the water. In a rare occurrence, the Garrison Diversion Conservancy District issued a report critical of the Sykeston Canal. The District supported the Mid-Dakota Reservoir over the Sykeston Canal because of lower costs and less risk of biota transfer (Kelly 1989). Following the 1986 Act, activities began on municipal, rural, and industrial projects; mitigation and wildlife habitats; and construction continued on some of the water delivery features. The continued evaluation of a smaller Lonetree Reservoir as a project feature and further analysis of the recommended Sykeston Canal deferred progress with construction of the principal water delivery facilities. President Bush did not include any funding for Garrison in his submitted fiscal-year 1991 budget.

In connection with the Administration's decision to terminate Garrison Diversion funding in fiscal-year 1991, the Secretary of the Interior established a task group to develop a policy on support for future funding of the authorized project. The task group's decision was to continue funding only those features of the reformulated project which were consistent with contemporary water needs, national priorities, and the history of Garrison Diversion, but not to fund features which would be used for irrigation. The recommendations also included

- continuation of the municipal, rural, and industrial grant program;
- Indian municipal, rural, and industrial water supply programs;
- irrigation development on 17,580 acres to include two Indian reservations;
- continued operation for the Oakes Test Area research activities;
- recreation; fish and wildlife mitigation and enhancement initiatives; and
- a minimum level of operation and management on the already constructed main supply system facilities.

Funding for these features would be considered by the administration within the context of national priorities. Funding for completion of the non-Indian project irrigation facilities and for related principal after supply works were completed would not be considered.

In November 1994, the North Dakota congressional delegation and Governor Schafer requested that the United States Bureau of Reclamation initiate a collaborative process to find a consensus plan that would meet the contemporary water development and stewardship needs of the state. The collaborative process included representatives of the Standing Rock Sioux, Spirit Lake Sioux, Three Affiliated Tribes, the congressional delegation offices, and the governor's office. The U.S. Bureau of Reclamation provided technical and administrative support. Under the guidance of the collaborative group, the Bureau began a series of studies for the water supply needs of the state. In 1995, the North Dakota Legislature repealed a portion of the North Dakota Century Code dealing with the preservation of wetlands. The National Wildlife Federation interpreted this action as withdrawal of State support for the Statement of Principles and withdrew from the collaborative process.

Garrison Diversion today

Work continues by the GDCD to prepare amendments to the Garrison Diversion Reformulation Act of 1986. GDCD proposes that a completion plan for Garrison Diversion include federal appropriations to complete a downsized principal water delivery system to provide reliable, high-quality water to areas of need in central and eastern North Dakota. The amendments to the legislation call for study of possible alternatives to meet the Red River Valley's need for water. Any feature that would ultimately deliver Missouri River water would comply with the 1909 Boundary Waters Treaty with Canada. Today, Garrison Diversion has taken the project one step further.

Canadian concerns

International problems arise from the fact that most of the return flows from the diversion of water will end up in Canada. The Canadians believe that this will have a harmful effect in Manitoba and Lake Winnipeg, because the water may be of poor quality and may cause pollution leading to the injury of health and property in Canada (Grosz and Leitch 1989). Canadians expressed the same fears as our own environmentalists about the adverse effects on fish and wildlife, and the unpredictable changes that may take place by diverting living organisms from the Missouri River system into the rivers flowing north. They are quite sure that Garrison Diversion will have adverse effects in Canada and will violate the Treaty of 1909. The Canadians have indicated that they will not agree to trade-offs in other problem areas; they do not accept the idea of compensating benefits through augmented stream flows; and they would refuse to accept monetary compensation for injury caused.

Deep commitments and lifetime careers have been devoted to the Garrison Diversion Project. It is believed by many people to be of great importance to the welfare of North Dakota. On the other hand, a senior Canadian official, Matthews, stated that he thinks "this seemingly minor problem could evolve into the most serious political issue between the two countries, and he hoped it would not become so ... it would be a terrible mistake to permit this emerging issue to develop into a major irritant in our relations." Awareness of the issues before it was too late has led to avoiding major political confrontation (Matthews 1974).

Canadian concerns over the GDU were expressed in 1970. The Canadian government, with the support of the Manitoba Naturalists Society, Canadian House of Commons, and the Canadian Department of External Affairs, requested a moratorium on the project in 1973. Construction was halted in 1974, until the Canadian objections could be satisfactorily addressed. Concerns about GDU as a potential medium for biota transfer were largely generalized and speculative in the project's early stages in the mid-1960s to the mid-1970s. The IJC identified some concerns largely based on expert opinion and literature review. Although the areas of concern have changed little since the mid-1960s (Grosz and Leitch 1989, and Sayler 1990), GDU opponents have focused their opposition. (Grosz and Leitch (1989) identified Canadian concerns regarding GDU in North Dakota (fig. 2.3)). Their objectives were to categorize Canadian concerns/issues regarding GDU to identify areas of potential research and to prepare research proposals for those priority areas where science could help ameliorate concerns.

The International Joint Commission with six commissioners, three appointed by each government, established the International Garrison Diversion Study Board. It produced studies indicating that construction and operation of the Garrison Diversion project would have an adverse effect on Manitoba, causing injury to health and property as a result of adverse impacts on the water quality and biological resources of the province.

In 1984, the Garrison Diversion Commission, a federal commission appointed by the Secretary of the Interior, recommended the focus of the project be shifted from irrigation to municipal water delivery, and that the controversial aspects of the project be deleted. All irrigation would be restricted to lands that drain, or could be drained, into the

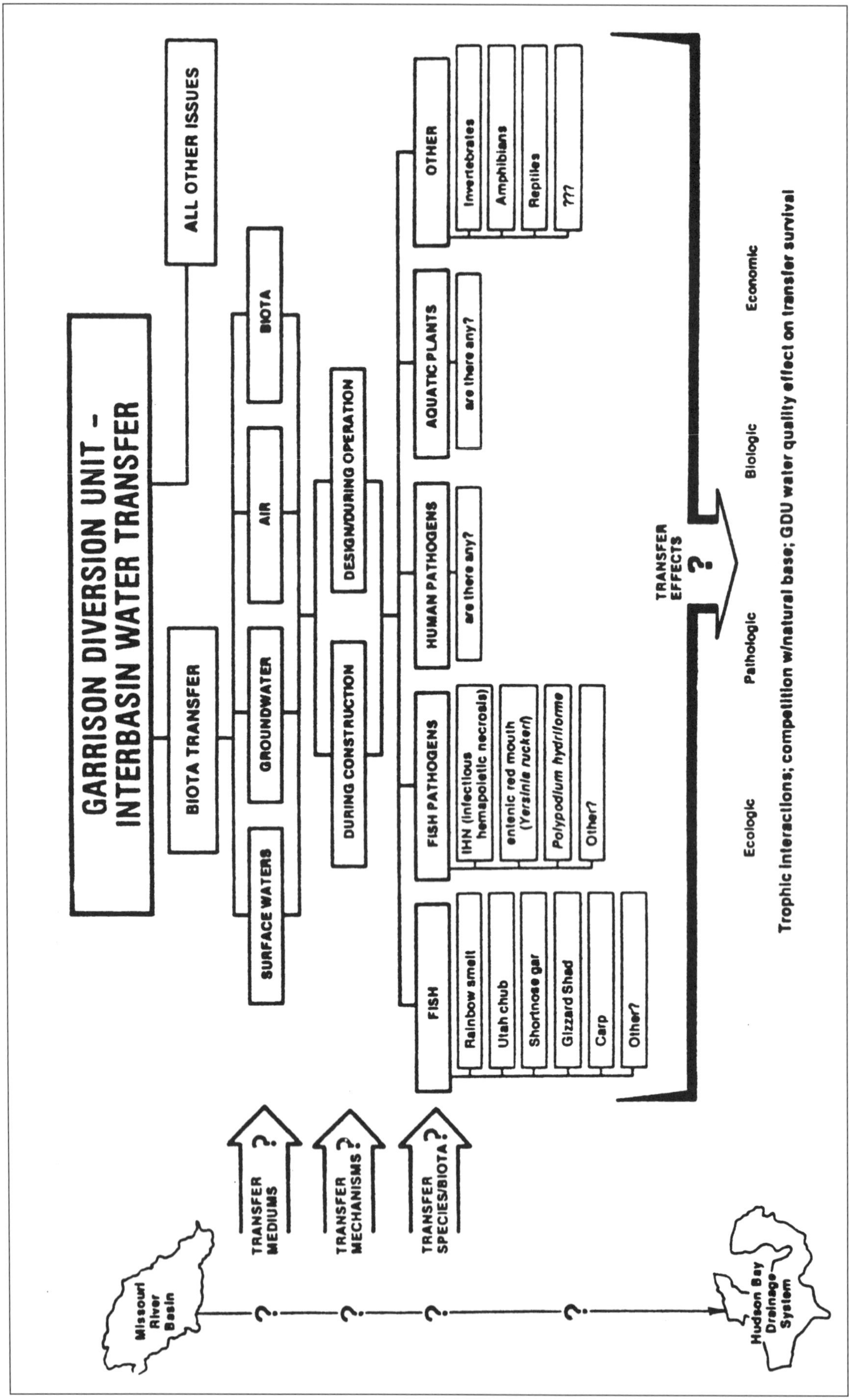

Figure 2.3 Biota transfer study components.

Missouri River basin. This amended version reduced irrigation acreage from 250,000 acres to 130,000, with emphasis on municipal, industrial, and rural domestic water supplies, and deleted the controversial aspects of the project (P.L. 98-360). The commission proposed studies and consultations with Canadian officials to develop methods that would permit the use of Missouri River water under future proposals for irrigation in areas of North Dakota that lie within the Hudson Bay drainage. The Garrison Diversion Reformulation Act of 1986 (P.L. 99-294) reauthorized the project, excluding all features that might adversely affect Canada. In response to the Reformulation Act, North Dakota's Governor wrote in 1986 to the state engineer: ... the possibility of ever obtaining Missouri River waters in eastern and northern North Dakota will depend on satisfying the Canadians and other opponents, as well as Congress, that Missouri River water can be safely transferred to and used in the Hudson Bay drainage.... We must marshal our scientific and political resources to meet the challenge in a careful, sound, and honorable manner that envisions many decades of effort.... It will require the cooperation and action of a number of state, federal, local, and private agencies and people, working over many years and probably several decades, to identify river basin water transfer problems and propose methods of mitigating or eliminating those problems.

Interbasin biota transfer study program

Research related to biota transfer from the Missouri River basin to the Hudson Bay drainage basin began in 1988. Former Governor William Guy was instrumental in establishment of the biota study program. The 1985 North Dakota Legislature established a mandate, under Senate Concurrent Resolution 4081, requesting the Governor to initiate a study of the possible adverse effects of transfer of fish species, biota, and pathogens between the Missouri River basin and the Hudson Bay drainage basin (Krenz 1994). Governor George Sinner established the Interbasin Biota Transfer Study Program (IBTSP) by executive order. A change of name to Interbasin Water Transfer Study Program was made by the Governor's Oversight Committee in October 1993 to reflect the broad objectives of the program.

The criteria for establishment of IBTSP were contained in the U.S. Congress and Public Law 98-360, which established the Garrison Diversion Unit Commission. The Commission's final report (dated December 20, 1984) called for a one-time baseline survey of species and pathogens in the Hudson Bay drainage basin, and it specified that the report should be the responsibility of an independent, international organization not affiliated with the Governments of the United States or Canada (Krenz 1994). Governor Sinner established two groups within IBTSP: the Governor's Oversight Committee and a Technical Advisory Team (TAT).

The Oversight Committee consisted of representatives of a broad cross section of governmental agencies and was responsible for designing the overall direction of IBTSP. Membership included the North Dakota State Engineer and representatives from the State Health Department, State Game and Fish Department, Garrison Diversion Conservancy District, U.S. Bureau of Reclamation, U.S. Fish and Wildlife Service, U.S. Environmental Protection Agency, North Dakota State University, University of North Dakota, and North Dakota Water Resources Research Institute. Former North Dakota Governor William Guy, who played a pivotal role in establishing IBTSP, was the first chairman.

Individuals named to the TAT were knowledgeable about Garrison Diversion and the issues surrounding it. Others were recognized experts in a variety of academic disciplines with limited knowledge of the proposed water transfer. Gene Krenz, Director of the State Water Commission's Division of Planning, was named as project coordinator at the state level. Jay A. Leitch, Professor of Agricultural Economics at North Dakota State University in Fargo, became the IBTSP director and the TAT chair. The purpose of TAT was to determine the specific research to be conducted, to judge the merits of proposals submitted in response to requests for scientific studies, and to award funds provided by the Garrison Diversion Conservancy District, the North Dakota State Water Commission, and the U.S. Bureau of Reclamation to project Principal Investigators (PIs).

Formation of the TAT accomplished several important objectives. Its membership was international, having representatives from both sides of the international boundary. It was independent of any governmental bodies and the agencies that funded it. TAT also became the body for directing interdisciplinary studies. The project was designed to address the dual and complementary purpose of answering questions, raised by the sponsors and others, and adding to the basic science of biota transfer. The underlying purpose was to help decision makers use science to address social and economic problems. Central to this purpose was researching aquatic biota in the context of interbasin water transfers.

Biota, broadly speaking, is all plant and animal life of a particular region. The overriding issue is the unintended transfer of undesirable "biota" from the Missouri basin to the Hudson Bay drainage basin (fig. 2.4). However, "biota" had never been explicitly identified nor defined. Biota transfer could occur in several ways, as depicted in a schematic of the range of issues, biota, and pathways (fig. 2.5) developed by Grosz and Leitch (1989). Interbasin water transfers and potential biota transfer can be local or global, or on a continuum between. The Garrison issue involves potential global transfers.

In a review of interbasin water transfers with specific attention to biota transfers, Padmanabhan, Jensen, and Leitch (1990) identified large-scale water transfers around the world, such as Orange-Great Fish River Transfer in South Africa, the McGregor Diversion in Canada, and Garrison Diversion in the United States. Very little information is available on the biota transfer aspects of these projects. A summary of interbasin water transfers is presented in Chapter 3.

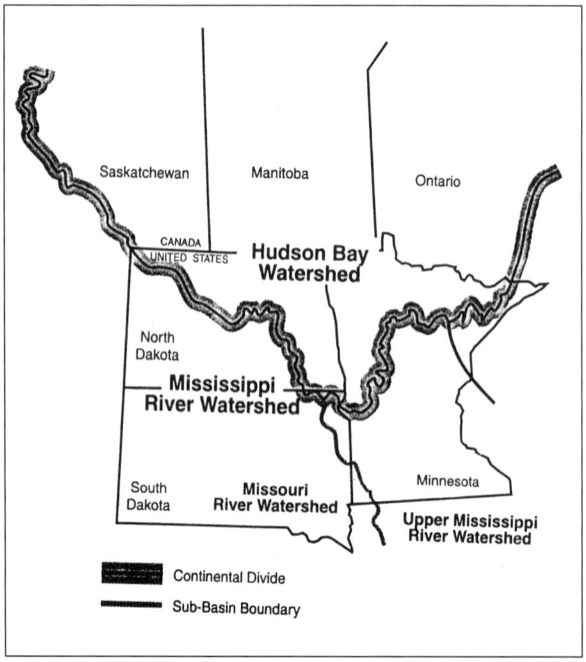

Figure 2.4 Continental Divide separating the Mississippi River and Hudson Bay Basins.

Staff of IBTSP under the auspices of the North Dakota Water Resources Research Institute at North Dakota State University provided the information transfer link between the politicians who wanted answers and the scientists who wanted to "do science" and remain apolitical. Scientists were ensured of working on technical, physical, biological, and other "hard science" issues, while remaining isolated from the politics of GDU. Staff brought together all interests and resources facilitating the collaborative research. The project director was supportive of the decision-making process and helped bridge the interpersonal aspects of international inter- and multidisciplinary research. Project sponsors (North Dakota State Water Commission, Garrison Diversion Conservancy District, and the U.S. Bureau of Reclamation) were assured good science, relevant research, and results usable in the policy arena. Other participants in this program included sponsor representatives, research administration personnel from each institution, and politicians. Each participant looked at the program differently. Each had different areas of expertise, different credentials, and different values. Incentives and risk varied among the participants.

A review of IBTSP annual reports, as of June 1996, reveals that 37 proposals had been received since 1988 with combined budgets exceeding $1,272,000. Twenty projects were funded, totaling approximately $790,000, with project duration from three months to three years. The awards ranged from $2,000 to $104,500 with $39,500 being the average award given.

A new manager of the Garrison Diversion Conservancy District (GDCD), Warren L. Jamison, was appointed in 1993. He proposed a revisioning of the Garrison Diversion project to overcome the barriers represented by Canadian objections to potential biota transfer and the United States' federal reluctance to allocate funds. The revisioning plan calls for repairs to restore the McClusky and New Rockford Canals and to build a pipeline connecting link between them. Wildlife mitigation includes the James River diversion facilities, consist-

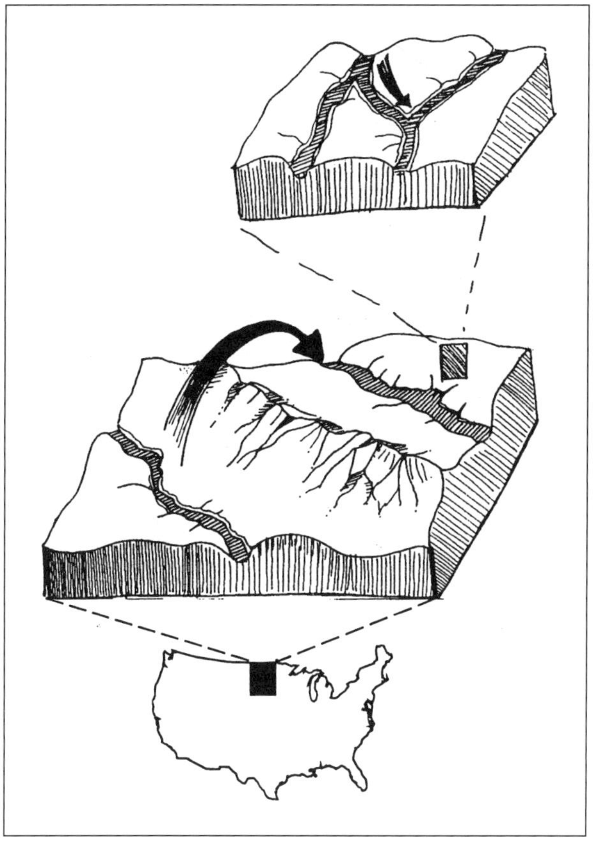

Figure 2.5 Pathways for Biota Transfer.

ing of a short feeder canal, the Arrowwood Refuge bypass facilities, and other facilities needed to control flows to the Sand Lake area. A pipeline to deliver water to Fargo and Grand Forks was also discussed.

Phase Two of the revisioned plan calls for the establishment of a trust fund to be funded by an annual federal appropriation equal to five percent of the Bureau of Reclamation's budget or $35 million annually for ten years or until $320 million is reached. The purpose of the trust fund would be to support regional economic development by providing loans and grants for Indian and non-Indian water supply development beyond the major supply works established in Phase One. A collaborative process was initiated by GDCD, whose purpose is to garner agreement on the new vision. Given that the proposal is a departure from the 1984 reauthorization act, a new consensus on the part of the political leaders is necessary.

This is where the Garrison Diversion Project stands today. It is radically different from the one originally envisioned. The reformulation bill attempted to address Canadian concerns, but the transfer of Missouri River water into the Hudson Bay drainage basin via the Red River remains controversial until some scientific solutions can be found.

References

Bureau of Reclamation. 1986. *Draft Supplemental Environmental Statement Commission Plan.*

Grosz, K. and J.A. Leitch. 1989. *Identification and Analysis of Canadian Concerns*, Report to North Dakota State Water Commission. Fargo: North Dakota State University, Water Resources Research Institute.

Hargreaves, M. W. 1957. *Dry Farming in the Northern Great Plains, 1900-1925.* Cambridge: Harvard University Press.

Harrison, H.H. August 1904. *The Missouri River and Dams, Maxwell's Talisman.* Chicago: Guild of the Talisman Workers.

Hart, H.C. 1957. *The Dark Missouri.* Madison: University of Wisconsin Press.

International Joint Commission of Canada and the United States (IJC). 1976. *Transboundary Implications of the Garrison Diversion Unit.*

International Joint Commission of Canada and the United States. 1977. *Transboundary Implications of the Garrison Diversion Unit.*

Josephy, A.M., Jr. March 1975. Dr. Strangelove builds a canal, *Audubon* 77(2):76.

Kelly, P.E. 1989. *Under the Ditch: Irrigation and the Garrison Diversion Controversy.* M.S. thesis. Fargo: North Dakota State University.

Krenz, G. 1994. Interbasin biota transfer study program: an overview and progress report. In Leitch, J.A. and D.J. Christensen (eds.), *1990 Interbasin Biota Transfer Study Program Proceedings, North Dakota Water Quality Symposium.* Fargo: North Dakota State University, North Dakota Water Resources Research Institute.

Matthews, H.F., Jr. 1974. *International River Problems: Three Examples.* Sixteenth Session - Senior Seminar in Foreign Policy, Department of State. pp. 14-28.)

Missouri Basin Inter-Agency Committee. June 1952. *Missouri River Basin Development Program.* Washington: GPO.

North Dakota Cooperative Extension Service. 1937. *Agricultural Engineering Annual Report, 1936-37.* Fargo: North Dakota State University, Institute for Regional Studies.

North Dakota Extension Service. 1956. *Annual Report, 1955-56.* Fargo: North Dakota State University.

Padmanahban, G., K. Jensen, and J.A. Leitch. 1990. A review of interbasin water transfers with specific attention to biota transfer. pp. 93-99 in Fitzgibbon, J.E. (Ed.), *Proceedings of the Symposium on International and Transboundary Water Resources Issues.* Toronto, Canada.

Richardson, E. 1973. *Dams, Parks and Politics: Resource Development and Preservation in the Truman-Eisenhower Era.* Lexington: University Press of Kentucky.

Ridgeway, M.E. 1955. *The Missouri Basin's Pick-Sloan Plan.* Urbana: The University of Illinois Press.

Robinson, E.B. 1966. *History of North Dakota.* Lincoln: University of Nebraska Press.

Sayler, R.D. 1990. Fish transfers between the Missouri River and Hudson Bay basins: a status report and analysis of biota transfer issues. In Leitch, J.A. and D.J. Christensen (eds.), *1990 Interbasin Biota Transfer Study Program Proceedings, North Dakota Water Quality Symposium.* Fargo: North Dakota State University, North Dakota Water Resources Research Institute.

Whittlesey, N.K. 1986. *Energy and Water Management in Western Irrigated Agriculture.* Boulder: Westview Press.

CHAPTER 3

A review of biota transfer aspects of interbasin water transfers

G. Padmanabhan, Mariah J. Tenamoc, David R. Givers, and Jay A. Leitch

Executive summary

Water transfers across basin boundaries have been taking place in many parts of the world for centuries. The volumes and route lengths of water transfer schemes have steadily increased. Technical feasibility of implementing large transfers is already evident and can only improve with the advance of technology. However, concern over transfer of biota through water transfers and the environmental implications has slowed down the increase in large scale water transfers. This chapter reviews the current knowledge base related to the biota transfer aspects of water transfers. While extensive literature can be found on economic and technical feasibility of water transfers, only a few case studies are available on biota transfer aspects. Many unresolved issues remain in evaluating biota transfer implications. New broad-based, multidisciplinary techniques need to be developed for evaluating biota transfer implications. Currently available techniques are inadequate. The amount and type of baseline data gathered on such projects are usually found inadequate to carry out meaningful evaluations. In order to make large scale transfers acceptable in the future, engineering solutions need to be found to minimize the transfer of biota. In the absence of effective biota control methods, interbasin water transfers may remain environmentally infeasible for a long time to come.

Introduction

Interbasin water transfer is defined as the process of withdrawing water from one river basin, transporting it across the basin boundary, and delivering it in another river basin. Transfers across watershed boundaries may occur through canals, channels, conduits, and tunnels. Water in these transfer schemes may move by gravity or may have to be pumped. Large-scale water transfer systems appear to have been constructed mostly in large countries with uneven natural water resources distribution. Examples are Australia, Canada, China, India, South Africa, the former Soviet Union (FSU), and the United States.

Thousands of water transfer systems are in operation currently. They differ in purpose, characteristics, technical difficulties, and scale. The water diverted is used for irrigation, water supply, navigation, and other purposes. In recent times, the water transfer systems have become more complicated because of the larger amounts of water transferred and the longer routes of transfer. The recently built diversion dams are longer and higher, resulting in larger reservoirs. Pumping stations capable of raising water hundreds of meters are part of present-day transfers. Consequently, the magnitude of the impact of water transfers on the environment has steadily increased over the years.

In characterizing a water transfer system, one may think in terms of local, zonal, interbasin, and interzonal. The local transfers occur within a river basin discharging into a lake or sea. The lengths of local transfer can reach up to a maximum of 100-150 kilometers. The zonal transfers occur within a physiographic zone. All the water taken from the river is used beyond the donor basin. These transfers are tens to hundreds of kilometers in length. The interbasin water transfers take place from a basin to an adjacent basin within one physiographic zone involving up to hundreds and even thousands of kilometers. The interzonal transfers connect river basins in different physiographic zones to different seas and oceans involving routes of thousands of kilometers. However, the term "interbasin water transfer" is used in the hydrologic literature to imply transfer of water between any two, or among any number of, river basins.

The amount of water involved and the length of route of transfers are the most important factors in characterizing the magnitude, complexity, cost, and the impact on the environment of a water transfer system. Therefore, a parameter combining the amount and the distance of transfer is used to classify water transfer systems. This parameter, "transfer rate factor," is defined as the product of the route length and the annual volume of water transferred, i.e., kilometer x $kilometre^3$/year, or $km.km^3.yr^{-1}$. The water transfers can be classified as minor, average, large, very large, and super large with the transfer rate factor ranging from up to 100 $km.km^3.yr^{-1}$ for minor transfer to those exceeding 10,000 $km.km^3.yr^{-1}$ for super large transfer. Considering some of the largest operating transfer systems and the projects being implemented and those being planned, Shiklomanov (1985) forecasted the largest capacity of such transfer systems to be between 18,000 and 22,000 $km.km^3.yr^{-1}$ by the year 2000.

The implementation of most of the water transfer systems used to be based on technical feasibility and economic justification. More recently, environmental considerations have assumed greater importance than the economic aspects, leading to slow down or cessation of work on new and proposed interbasin water transfer projects. With increased awareness, special interest groups question the social acceptability of these projects, which then become incorporated into the political process. A major environmental concern associated with large-scale water transfers is the transfer of biota—fish, mollusks, insects, bacteria, viruses, and parasites. With the recent emergence of issues about biota transfer, it became necessary to review literature to discover what, if anything, was already known about biota transfer across drainage basins.

An interbasin water transfer project may impact physical, biological, and human systems of the water exporting region (donor), the transfer region (conveyance), and the importing region (receiving) (Seagel and Adam 1986; Reynolds 1986; Ganging and Zhang 1986; Yuexian and Jialian 1983). Physically, water transfers may affect the quantity of available water, water level, discharge, and velocity. The quality of water may be affected also with changes in sediments, nutrients, turbidity, salinity, alkalinity, temperature, and toxic chemicals. There are implications not only for water, but for land as well. Sedimentation, salinity, alkalinity, waterlogging, and earthquake inducement may change as a result of water transfer. The atmosphere, as well, may exhibit changes with temperature, evapotranspiration, and changes in micro-climate.

Human systems are impacted by water transfers through production in agriculture, hydropower, navigation, and recreation. Some socio-cultural aspects of the human system which are impacted by water transfers involve the social costs, infrastructure developments, anthropological effects, and political implications (Biswas 1983).

Although the physical and human systems are of interest, and the environmental impact of interbasin water transfer on them are extensive, the focus of this review is on the biological system with respect to aquatic changes that occur as a result of water transfer across drainage basins. Changes may occur in fish and aquatic vertebrates, aquatic plants, and in the type of aquatic diseases. Animals and vegetation may be impacted by water transfer effects on habitats, either through loss of habitat or enhancement of habitat (Biswas 1983). Faunal transport and changes in water quality and quantity also occur with effects on aquatic and terrestrial ecosystems.

One immediate effect in the exporting region, if a diversion dam is constructed, is flooding of the impoundment area, sometimes inundating prime farmland, wildlife habitats, forest resources, fisheries, and other productive resources. A change in surface and ground runoff and a rise in water table leading to secondary salinization of soil also may result (Yuexian and Jialian 1983). In addition, conditions below the dam can change. The structure acts as a barrier to fish movement. Changes in water flow, temperature, dissolved solids, turbidity, total alkalinity, and pH may lead to changes in aquatic habitats. Flow rate reduction may change hydrological regimes and alter the physical and biological morphology of the exporting region (Seagel and Adam 1986; Reynolds 1986).

Studies have been conducted to determine the effects of changes in flow on wildlife and fisheries (O'Keefe and DeMoor 1988; Meador 1992; Petitjean and Davies 1986; Coutant 1985; Thomas and Box 1969; Seagel and Adam 1986). Changes in river regime and morphology might eliminate riverine habitat for wildlife, but some recovery might occur when the river stabilizes. For example, studies on the McGregor River diversion in Canada show that these impacts can be realistically expected (Seagel and Adam 1986). Fish populations dependent on the mainstream would be affected during the transition phase, when major reductions of game fish could occur due to losses of rearing, feeding, and spawning habitat and increased stress (Seagel and Adam 1986).

There may be effects also in the receiving region. Reduction in seasonal variation of flow is one example as evident in the Great-Fish River transfer in South Africa (O'Keefe and DeMoor 1988). Interbasin transfers can cause significant changes in an ecosystem. Meador (1992) classified potential impacts into four groups. First, introduction of non-indigenous aquatic organisms may cause significant changes in the ecosystem. Aquatic organisms include parasites, plants, invertebrates, and fish. Examples of 'introduction' can be found in the republics of the former Soviet Union (Gurvich, Novikov, and Saifutdinov 1975), in Great Ouse transfer in Great Britain (Guiver 1976), and in Orange-Great Fish River transfer in South Africa (Laurenson and Hocutt 1985). Second, changes in flow may cause significant changes in the ecosystem. For instance, increases in flow in the upper reaches on the order of 500-800 percent have caused shifts in benthic invertebrate communities of the Great Fish River, South Africa (O'Keefe and DeMoor 1988). Third, alteration of volume of flow and timing of discharge can substantially impact aquatic habitats. And, finally, interbasin transfers may result in changes in water quality, such as conductivity, turbidity, alkalinity, and pH, in the receiving basin. These changes, in turn, may affect the spawning of native fish species. This was evident in the Murray and Murrumbidgee Rivers receiving transferred water from Snowy and Eucumbene Rivers in Australia (Meador 1992; Thomas and Box 1969).

Transfers to nearby basins with similar fish and habitats would have minimum effects. The effects can vary from negligible to large depending on the transfer project. Fish transferred in an interbasin water transfer can cause problems through rapid population growth and possible deleterious interactions with native fish in the receiving basin. By competing with indigenous fish for food and space, a transferred fish population can lead to disruption in the native fisheries (International Garrison Diversion Study Board 1976; Owen and Elsen 1976; Loch et al. 1979).

Transferred species may not interact immediately with an ecosystem. Rather, a complex set of events may occur over time. Thus, an introduced species may or may not become a permanent inhabitant. The ability to adapt and survive will determine the long-term effects of species introduction on the receiving waters. In some cases, introduced fish are harmful because of predatory and competitive effects on native fish. In other instances, a species may have no measurable or observable effect. Introduced species may have a negative economic impact if they displace or alter an existing fishery and have little value as game fish.

Two biologic concerns of major importance, before a species is introduced, are whether it will survive in the new environment and whether it will feed on or be preyed upon by other species of direct interest. Survival and success of an introduced species is a function of intrinsic and extrinsic factors. For example, genotype of the species (intrinsic) interacts with the environment (extrinsic) to determine the success of an introduced species in its transferred environment. It is usually possible to determine from collected data how effective a species has been at colonization of new regions. Impacts of certain species may be predicted based on ecological modeling of invasion of closely related species. Careful examination should include an evaluation of possible impacts on the communities into which they might spread and considerations for improving conditions for native species present.

World-wide concern is focused on characteristics of invasive biota, on the impact and management of altered ecosystems (SCOPE 1983; Shafland and Lewis 1984; Seagel 1987). Planning and managing transfer projects require scientific information on biota transfer. Management of aquatic ecosystems impacted by invasive biota requires knowledge and understanding of the behavior of introduced species in their new environments. Viral, bacterial, and parasitic transfers can also take place, along with the transfer of fish between basins. The study of aquatic fauna, especially fish species, their distribution, life cycle, habitat types and response to physical changes, is necessary in ecosystem studies. Knowledge of viral, bacterial, and parasitic species present in donor and receiving waters, their life cycle, their history, and their pathogenicity is equally important. Available information, in general, is inadequate for comprehensive assessment of impacts and risks. Studies on "introduced species" (species deliberately or inadvertently transferred to an area outside its native range) are needed for understanding the potential effects of biota transfer. To date, comprehensive case studies are almost nonexistent because:

- the pre- and post-monitoring of biota transfer and impacts must be over long periods of time to draw meaningful conclusions,
- there are not enough baseline studies, and
- the science of evaluating biota transfer implications is in its infancy.

Ecological costs were not included in the project costs of earlier water transfer projects which were easily justified economically. However, there is an increasingly compelling need to include the ecological costs in all future projects. Since 1969, environmental impact statements have been required for all transfer projects in the U.S.A. The larger a project, the more difficult it is to estimate the ecological cost with any degree of certainty. Although it is understandable that projects beyond a certain scale will have effects on the environment, and costs that are not easily justifiable, quantitative evaluation of these projects is not tractable.

For reliably calculating this cost, we need the following evaluations.

- A comprehensive evaluation of the effect of the project on the environment over the entire area of the project—donor, receiving, and conveyance regions.
- Evaluation of the effects of environmental changes on the economy and social conditions of regions.
- Evaluation of a variety of measures to minimize the possible undesirable effects.

The first one is the most difficult. The ecological cost of transfer systems depends on the type, volume, and length of transfer. It would be minimum for local transfers and maximum for large transfer systems involving large areas with varied relief.

Many large-volume interbasin water transfers exist in countries around the world. Large-scale water transfer systems have been constructed mostly in large countries with uneven spatial distributions of water resources. Following are brief descriptions of selected interbasin water transfer systems in the United States, Canada, the FSU, South Africa, China, Australia, and India.

United States

In western U.S.A., the Colorado River basin is the donor basin for waters transferred to the Great Basin, the Pacific coast, the Upper Missouri basin, Arkansas, and the western area of the Gulf of Mexico basin. About 23 km^3./yr of the discharge are used in the local basin and adjacent basins. About 7 km^3./yr of water are diverted beyond to California. There are over 16 major diversion projects in the western states alone, excluding numerous projects in California. The California aqueduct system delivers water from northern California rivers to the deltas of Sacramento and San Joaquin — a transfer of 5.2 km^3./yr over a distance of 300 miles. All of the U.S. systems of interbasin transfer redistribute about 30-35 km^3./yr of water. In Texas, diversions from the Canadian River serve a population of 450,000 in towns and cities in the Red, Brazos, and Colorado River basins (Templer and Urban 1995). Comprehensive consideration of the long-term (to the year 2020) prospects for water in the United States began in the 1960s. It was established then that the United States would soon face great problems of fresh water supply. One way in which the water supply problems could be addressed was by large-scale redistribution of water resources. Many plans have been proposed since the 1960s. Interest deteriorated during the 1970s because of:

- a slackening in the demand to increase the irrigated areas in arid regions and the growth in agricultural production in the humid ones,
- re-use of water by industry and use of saline waters for some purposes,

- problems of sharing water between the U.S.A. and Canada and between different States in the U.S.A.,
- environmental implications of water transfers, and
- the engineering difficulties and the large expenditures likely to be encountered (Biswas 1978, Howe and Easter 1971, Micklin 1977).

Physiography in the U.S.A. is also unfavorable to construction of large-scale water transfer systems. The systems would generally involve canals, as many as thousands of kilometers long, and involve lifting water over 1,000 meters. There had been little or no attempt when designing water transfer systems in the U.S. prior to the 1970s to consider the impact on the environment (Shiklomanov 1985). Another diversion project, "Garrison Diversion," has been the subject of many biota transfer studies and engineering solutions to address those implications. This project is described later in this chapter.

Canada

An excellent review of the water diversions in Canada can be found in Day and Quinn (1992). They reported case studies of five diversion projects from earliest to recently implemented and also, at the same time, from smallest to largest in scale — the Long Lake and Ogoki diversions in northern Ontario, the Nechako-Kemano diversion in central British Columbia, the Churchill-Nelson diversion in northern Manitoba, and the James Bay diversion in Quebec. They also reported the impact on fish, erosion, timber, parks, pollution, and reservoir and spillway area residents. Economic effects and cost-benefit analysis were also analyzed for each case study. Overall, it appears that biota transfer impacts have not been addressed adequately in these projects.

There are no similar interbasin water transfer projects anywhere in the world which divert as much water as those in Canada (Shiklomanov 1985). The transfer systems in Canada are large by the volume transferred, but shorter in length of routes of transfer compared to other countries. The volume transferred in Canada is about 140 km^3./yr. Projects which divert exceptionally large volumes of water and have been constructed during the 1970s include the Churchill Project in Manitoba (24 km^3./yr), the James Bay Projects in Quebec (52.5 km^3./yr), and the Churchill Falls Project in Newfoundland (17 km^3./yr). According to Shiklomanov (1985), there are unique features to Canadian transfer systems. They are characterized by large volumes of water and involve short man-made routes along which flow occurs by gravity. No large-scale pumping is necessary. Shiklomanov states (1985, p. 358), "The construction of transfer systems in Canada is favored not only by large amounts of surface water but also by the undulating topography produced by the Quaternary glaciation. The dense random network of river valleys and ancient glacier beds is not always filled with water, and there are a great number located at different elevations." Transfers are made easier because the divides are usually low and intra fluvial areas are narrow (Timashev 1982), and the lakes often serve as reservoirs to provide control. Transfer systems in Canada have usually been constructed where there has been little development, and their impact on the environment has been small.

Much attention has been paid, especially in recent years, to ecological problems in transfer areas (Baxter and Glaude 1980; Environment Canada 1975; Soucy 1978; Clark et al. 1978). Issues also include technical, economic, ecological, social, and political problems.

A program of studies was initiated in British Columbia, Canada, in 1976, to determine environmental effects of a proposal to divert waters of the McGregor River (Pacific drainage) into the Parsnip River (Arctic drainage). Extensive faunal transfer studies were conducted. Concerns focused on fish, viruses, bacteria, and parasites. No potentially pathogenic bacteria or viruses were detected in any of the fish species. Three forms of parasites were identified in the Pacific drainage which were not found in the Arctic drainage (Parsnip River) and are expected to pose the greatest threat to the fisheries resource. Neither total fish populations nor prevalence of disease can be determined with any degree of confidence. Difficulties remain in analyzing and interpreting studies of this nature (Seagel 1987).

Lack of baseline studies before projects are implemented makes it difficult to measure changes in parameters and to conduct effective monitoring programs (Peet and Day 1980; Shiklomanov 1985; Reynolds 1986; Seagel and Adam 1986; Nicholaichuk and Quinn). However, studies continue to be conducted both on completed diversion projects and on proposed projects. More recent studies include aquatic-biota transfer issues.

FSU

In Russia, mid- and large-size water transfer systems were initiated early in the century. The total capacity of water transfer systems in the FSU is 60 km^3./yr. The Volga-Moscova Canal, taking water from the Volga basin to the Moskova basin, (completed in 1937) provides water to Moscow. The Great Fergana Canal in Central Asia was one of the largest canals (by the capacity and length) constructed at the time (1939). Water is transferred from the Naryn basin to the Kardarya basin. This system involves a transfer of 11 km^3./yr over a distance of 1100 km.

At present, the largest water transfer system in the world is the Karakum Canal, which diverts water from the Amudarya basin to the Ashkhabad basin and to areas of the Karakum desert. The FSU placed great emphasis on forecasting and evaluating impacts of large-scale projects on the hydrometeorological conditions and on the environment (Shiklomanov 1985). Multi-purpose research programs have been designed to evaluate the current state of fresh waters, to predict future changes in the water regime and on the environment in the absence of large-scale water transfers, and to predict the effects of water transfer projects on the environment. In a comprehensive evaluation of planned water transfers on the environment, scientists from more than 100 research institutions are studying such topics as:

- forecasting possible changes in the water regime and water balance of river basins downstream of the sites where the water is diverted,
- hydrological processes along the routes of transfers and in adjacent areas,
- hydrometeorological regime and their changes in the areas to which the water is diverted together with the effects on the regimes in inland water bodies, and
- possible changes in the climate and water circulation (Shiklomanov 1985).

South Africa

Water from the Orange River in South Africa is diverted to the Great Fish River. Studies of biota transfer in connection with the Orange-Great Fish River water transfer project commenced in 1975. This is one of the few transfer schemes that includes pre- and post-transfer studies. Interest and requests for biota studies resulted from the concern over the potential alteration of ecosystems. Studies conducted include hydrological and water quality changes, invertebrate changes, fish dispersal, and post-impoundment trends in the reservoir used for diversion. Data compiled on hydrology, chemistry, and invertebrates prior to the opening of the Orange-Fish Tunnel are compared with the data for post-transfer conditions (Laurenson and Hocutt 1985).

Flow patterns obviously altered after the transfer was implemented. Upper river flows in the Great Fish River changed from seasonal to perennial, while seasonal flows in the lower river were reduced. Inflow of low salinity water from the exporting (Orange River) region diluted the highly mineralized Great Fish River, thereby lowering the total dissolved solids. Invertebrate communities in riffles in the Great Fish River changed substantially as a result of the transfer with only one-third of taxa common to both pre- and post-surveys. Dominant species changed, but overall densities were not altered. Major changes in species can be attributed to the more permanent flow and increased area of erosional habitats (Guiver 1976; Hamman 1980; Cambray and Jubb 1977).

Fish species passing through the tunnel are monitored, and the possibility of their becoming established in the Great Fish River exists. It is clear from these studies that five Orange River species have an artificial access to the Great Fish River (Laurenson and Hocutt 1985). Biota transfer investigations of the Great Fish River are still considered to be in early stages.

China

In China, the 1,500 km. long Yun Ho (Grand Canal) crosses the Great Plain and connects the Yangtze, Huaihe, Huang Ho (Hwang-Hae), and Haihe rivers. Future projects include transfer of water from the Yangtze for irrigation and water supply in the North China Plain. Exporting, importing, and transfer regions are studied for possible environmental impacts. The studies include soil salinization, water quality and its management and protection, health problems caused by waterborne diseases, impact on aquatic environments, seepage problems from canals, water interception along drainage divides, sedimentation in canals and exporting regions, estuary problems in exporting regions, impacts on neighboring terrestrial and freshwater ecosystems, rising groundwater problems, inundation of land for reservoirs, microclimates, fisheries, water quantity, navigation and saltwater intrusions (Changming and Dakang 1983; Jinghua and Yongke 1983; Ganging and Zhang 1986; Shouquan 1983; Bangyi and Chunhuai 1983; Yuexian and Jialian 1983; Xuefang 1983; Changming 1985).

Australia

The major transfer system in Australia is the Snowy Mountain transfer system which carries water from Snowy River across the Great-Dividing Range to the Murray basin and to the Murrumbidgee, a tributary of the Murray River. The system provides 1.39 km^3./yr to the Murrumbidgee River and 1.00 km^3./yr to the Murray River. The transfer involves 80 km. of pipelines and 140 km. of tunnels.

India

India, characterized by high temperatures throughout the year, has an extremely uneven distribution of precipitation and streamflow in time and space (Shiklomanov 1985). Long-distance transfer of huge volumes of water has been practiced in India for over five centuries. A more intense development of water transfer has occurred over the last 30-40 years. At present, the main transfers are diverted from the three main tributaries of the Indus River (Ravi, Beas, and Sutlej) to the donor areas in the south of Punjab, Haryana, Rajasthan, and Jammu-Kashmir states. Murthy (1978) reported two large transfer systems for irrigation: the Rajasthan Canal project, involving a transfer of 16.5 km^3./yr over 180 km. from the Himalayas to the deserts of Rajasthan, and the Sarda Sahayak project, involving 15.4 km^3./yr over 260 km. from the Ganges River to the plains of the Ganges. The other completed transfer projects are Periyar diversion scheme, Parambikulam-Aliyas project, Karnool-Cuddapah Canal, Beas-Sutlej link, and Ramaganga diversion. Venugopalan (1979) reported the need for the creation of a single national management system diverting water from northern rivers of India to dry areas. One of the proposals under this concept will enable the development of a national water network. However, the implementation of the proposal to build a national water network is difficult because of colossal investment and technical considerations. The transfer will involve 22 x 10^9m^3. of water to be lifted over 400 m. height from the Ganges River, 16 x 10^9m^3. over 15 m. over the Brahmaputra River, and a transfer of 50 x 10^9m^3. from the west-flowing rivers to the province of Gujrath. Another proposal is to intersect the entire continental drainage using two contour canals at commanding elevations — one along the foothills of the Himalayan Mountains and the other along the Brahmaputra River on the southern plateau of India. Although environmental impacts were of no concern in earlier water transfer projects; recently, several of these projects, even smaller in magnitude, have run into difficulties due to objections on the basis of environmental impact. Biota transfer studies have not been undertaken in the implementation of transfer schemes.

Garrison Diversion, United States

The water transfer from the Missouri River basin in North Dakota, U.S.A., to the Hudson Bay basin in Canada, with the implementation of the Garrison Diversion, is the focus of studies reported in this book. Garrison Diversion is expected to provide irrigation; municipal and industrial water supply; and recreational, fish, and wildlife opportunities in North Dakota by transferring water from Missouri River basin to the Hudson Bay drainage basin. Completion of this transfer scheme would provide a direct connection between the Missouri River basin and the Hudson Bay drainage basin through the McClusky Canal (International Garrison Diversion Study Board 1976; Pearson 1983; Carroll and Logan 1980). Most of the drainage and waste waters would flow into transboundary streams.

Different fish faunas exist in the two basins (Loch et al. 1979; Clambey et al. 1983). Introduction of foreign or exotic species of fish, fish diseases, and fish parasites and other aquatic biota into Manitoba waters (Hudson Bay drainage basin) is a main concern (Grosz and Leitch 1989; Clarkson 1992). Biota transfer could create irreversible impacts on existing fisheries and aquatic systems, thereby leading to changes in commercial and recreational fishing.

Large rivers altered by human activity are prone to establishment of introduced species. Diversity of habitats in large watersheds raises the likelihood of successful exploitation by invading species. Introduction usually occurs with major environmental changes or other stresses on the native fish community. The Garrison Diversion Unit's features include both diverse habitats and ecosystem disturbances.

Detailed reviews of the life history and environmental requirements have been made to assess the probability that transferred species would become established (International Garrison Diversion Study Board 1976). Adverse impacts of biota transfer would be irreversible and would become evident in only about ten years, with the introduced species' full impact not being felt for 25 to 50 years after the completion of the project. Exotic fish introductions which have a high reproductive potential might successfully compete for food and space, replace indigenous forage fish, alter the balance between existing predators and their prey, carry parasites, and destroy some of the valuable present species. Also, there is a danger of fish disease being transferred from the Missouri River system. Impacts could create a general ecosystem destabilization, resulting in a population decline of important commercial and recreational fish (Pearson 1983; Carroll and Logan 1980; International Joint Commission 1977).

Twenty-one species of fish presently found in the Missouri system, but not in the Hudson Bay basin, may be translocated as a result of the Garrison Diversion Unit. Species that could create adverse impacts on indigenous fish resources are pallid sturgeon *Scaphirhynchus albus*, shovelnose sturgeon *Scaphirhynchus platyrhynchus*, paddlefish *Polyodon spathula*, shortnose gar *Lepisosteus platostomus*, gizzard shad *Dorosoma cepedianum*, Utah chub *Gila atraria*, smallmouth buffalo *Ictobus bubalus*, and river carpsucker *Carpiodes carpio*. Two fish diseases likely to be transmitted from the Missouri River drainage to the Hudson Bay drainage as a result of the interbasin transfer of water are identified as infectious hemopoietic viral necrosis (IHVN) and enteric redmouth (ERM), a bacterial disease. The potential for other parasites and fish diseases to be introduced appears to be low. The chance that the Garrison Diversion Unit will result in the introduction of plant diseases, terrestrial noxious weeds, or aquatic plants into Manitoba is low (International Garrison Diversion Study Board 1976).

Assessing impacts of future biota transfer is difficult due to lack of baseline data and lack of scientific techniques for forecasting future scenarios that would result from introduction of aquatic species. Detailed monitoring and baseline data collection over long periods are necessary to produce meaningful conclusions about future implications. Biota transfer issues continued to be raised and addressed in investigations on modifications of the Garrison Diversion Unit (Garrison Diversion Unit 1987). This is evidenced in the following chapters devoted to the systematic study of biota transfer issues as they relate to interbasin water transfer via Garrison Diversion Unit.

Summary of review

Extensive literature is available on economic and technical feasibility of water transfer schemes, but studies on biota transfer are relatively few. Only a few specific case studies are found in the literature. These include the Orange-Great Fish River Transfer in South Africa, pre-diversion studies of the McGregor Diversion in Canada, and pre-diversion studies of Garrison Diversion in the United States. Comprehensive case studies are nearly non-existent: there are not enough baseline studies; pre- and post-monitoring of biota transfer and impacts have to be conducted over long periods of time to draw meaningful conclusions; and the science of evaluating biota transfer implications is in its infancy.

Biota transfer concerns have received much attention only recently. There are many unresolved issues in evaluating biota transfer implications. Evaluation of biota-transfer implications and forecasting future scenarios of aquatic ecosystems are extremely complex. The techniques and information presently available to address biota transfer are inadequate. Broad-based, multi-disciplinary research programs are needed to adequately evaluate potential ecological impacts of water transfers. The program should include disciplines such as hydrology, limnology, invertebrate ecology, fisheries biology, botany, ecological modeling, and water resources engineering. Hydrologists, for instance, are needed to study water quality, erosion, sedimentation, and general hydrologic implications for land use as affected by interbasin transfers of water. Limnologists (biological, chemical) are needed to study water quality parameters such as nutrients, turbidity, salinity, and alkalinity in reservoirs. Spatial-temporal variability of invertebrates, introduction of invertebrates, and invertebrate-habitat relationships are some critical issues to be studied by invertebrate ecologists. Botanists are needed to study the introduction of non-indigenous aquatic and terrestrial riparian vegetation. Fisheries biologists are needed to study spatial and temporal variability of fish, effects of fish species introductions on native fauna, ichtyoplankton entrainment, fish-habitat relationships, and fish diseases. And systems analyst/modelers are needed to research the environmental responses under varying scenarios to better evaluate effects of various discharge rates on flora, fauna, and hydrology (Meador 1992). Review of the literature also reveals the need to develop engineering solutions to prevent or minimize biota transfer in water transfer schemes. In the absence of biota control methods, interbasin water transfers may remain environmentally infeasible for a long time to come.

References

Bangyi, Y. and C. Chunhuai. 1983. Some aspects of the necessity and feasibility of China's proposed south-to-north water transfer. In *Long Distance Water Transfer: A Chinese Case Study and International Experiences*, ed. Biswas, A.K. 321-332. Dublin: Tycooly International Publishing Ltd.

Baxter, R.M. and P. Glaude. 1980. Environmental Effects of Dams and Impoundments in Canada: Experience and Prospects. *Canadian Bulletin of Fisheries and Aquatic Sciences*, 205:34 pp.

Biswas, A.K. 1978. North American water transfers. An overview. In *Proceedings of the Task Force Meeting: Inter-regional Water Transfers Projects and Problems*, eds. Golubev, G., and A. Biswas, 79-90. Oxford: Pergamon Press.

Biswas, A.K. 1983. Long-distance water transfer: problems and prospects, In *Long-Distance Water Transfer: A Chinese Case Study and International Experiences*, ed. Biswas, A.K. 1-14. Dublin: Tycooly International Publishing Ltd.

Cambray, J.A. and R.A. Jubb. 1977. Dispersal of fishes via the Orange-Fish Tunnel, South Africa. *Journal of the Limnological Society South Africa* 3(1):33-35.

Carroll, J.E. and R.M. Logan. 1980. *The Garrison Diversion Unit.* Montreal: C.D. Howe Research Institute and National Planning Association.

Changming, L. and Z. Dakang. 1983. Impact of south to north water transfer upon the natural environment. In *Long-Distance Water Transfer: A Chinese Case Study and International Experiences*, ed. Biswas, A.K. 169-180. Dublin: Tycooly International Publishing Ltd.

Changming, L. 1985. Water transfer in China: The East Route project. In *Large-Scale Water Transfers: Emerging Environmental and Social Issues*, 103-118. Riverton, N.J.: Tycooly Publishing.

Clambey, G., H.L. Holloway, J.B. Owen, and J.J. Peterka. 1983. *Potential Transfer of Aquatic Biota Between Drainage Systems Having No Natural Flow Connection.* Final Project Report, Fargo, N.D.: Tri-College University Center for Environmental Studies.

Clark, R.H., H.B. Rosenberg, and F.G. Quinn. 1978. *Water Transfers in Canada*, 33 pp. Ottawa: Inland Waters Directorate, Department of Fisheries and the Environment.

Clarkson, R.N. 1992. *Canadian Concerns Regarding the Garrison Diversion Unit.* Third Biennial North Dakota Water Quality Symposium, Bismarck, N.D.

Coutant, C. 1985. Striped bass, temperature, and dissolved oxygens: A speculative hypothesis for environmental risk. *Transactions of American Fisheries Society* 114:31-61.

Day, J.C. and F. Quinn. 1992. *Water Diversion and Export: Learning from Canadian Experience.* Department of Geography Publication Series Number 36, University of Waterloo, Canadian Association of Geographers Public Issues Committee, No. 1.

Environment Canada. 1975. *James Bay Hydroelectric Project: A Statement of Environmental Concerns and Recommendations for Protection and Environment Measures*, 46 pp. Environment Canada.

Ganging, S., Sr., and L. Zhang. 1986. Environmental problems of the interbasin water transfer in China. In *Proceedings of IWRA Seminar on Interbasin Water Transfer*, 87-94. Beijing.

Garrison Diversion Unit. 1987. *Draft - Scope of Study, Studies Required by the Garrison Diversion Unit Reformulation Act of 1986.* Bismarck, N.Dak.: United States Department of the Interior.

Garrison Diversion Unit. 1987. *Report on Garrison Diversion*, Bismarck, N.Dak.: United States Department of the Interior.

Grosz, K.L. and J.A. Leitch. 1989. *Identification and Analysis of Canadian Concerns*, Report to North Dakota State Water Commission. Fargo: North Dakota State University, Water Resources Research Institute.

Guiver, K. 1976. Implications of large-scale water transfers in the UK: the Ely Ouse to Essex transfer scheme. *Chem. Ind.* 4:132-135.

Gurvich, L.S., Y.V. Novikov, and M.M. Saifutdinov. 1975. Study of sanitary problems in connection with interbasin transfer of river runoff. *Gig. Sanit.* 12:62-65.

Hamman, K.C.D. 1980. Post impoundment trends in the fish populations of the Hendrik Verwoerd Dam, South Africa. *Journal of the Limnological Society South Africa* 6(2):101-108.

Hermann, R. 1983. Environmental implications of water transfer. In *Long-Distance Water Transfer: A Chinese Case Study and International Experiences*, ed. Biswas, A.K. 159-168. Dublin: Tycooly International Publishing Ltd.

Howe, C.W. and K.W. Easter. 1971. *Interbasin Transfers of Water - Economic Issues and Impacts.* Baltimore: Johns Hopkins University Press.

International Garrison Diversion Study Board. 1976. *Report to International Joint Commission.* Report and Appendix C. Canada and the United States.

International Joint Commission. 1977. *Transboundary Implications of the Garrison Diversion Unit, Canada and the United States.* Bismarck, N.Dak.

Jinghua, W. and L. Yongke. 1983. An investigation of the water quality and pollution in the rivers of the proposed water transfer region. In *Long-Distance Water Transfer: A Chinese Case Study and International Experiences*, ed. Biswas, A.K. 361-372. Dublin: Tycooly International Publishing Ltd.

Laurenson, L.B.J. and C.H. Hocutt. 1985. Colonization theory and invasive biota: the Great Fish River, a case history. *Environmental Monitoring and Assessment* 6:71-90.

Loch, J.S., A.J. Derksen, M.E. Hora, and R.B. Oetting. 1979. *Potential Effects on Exotic Fishes on Manitoba: An Impact Assessment of the Garrison Diversion Unit.* Canada Fisheries and Marine Service Technical Report No. 838.

Meador, M.R. 1992. Interbasin water transfer: ecological concerns. *Fisheries* 17:17-22.

Micklin, P.P. 1977. A preliminary systems analysis at impact of proposed Soviet river diversion on Arctic Sea ice. *EOS Transactions, Amer. Geophys. Union* 62(19):489-493.

Murthy, R.S.S. 1978. Interregional water transfers: case study of India. In *Proceedings of the Task Force Meeting: Interregional Water Transfers: Projects and Problems*, eds. Golubev, G., and A. Biswas, 117-125. Dublin: Tycooly International Publishing, Ltd.

Nicholaichuk, W. and F. Quinn (eds.). 1987. *Proceedings of the Symposium on Interbasin Transfer of Water: Impacts and Research for Canada.* Saskatoon, Sask.: Canadian Water Resources Association.

O'Keefe, J.H. and DeMoor, F.C. 1988. Changes in the physico-chemistry and benthic invertebrates of the Great Fish River, South Africa, following the interbasin transfer of water regulated rivers. *Research and Management* 2(4):1-6.

Owen, J.B. and D.S. Elsen. 1976. *Effects of the Garrison Diversion Unit on Distribution of Fishes in North Dakota, Report No. 4: James River.* Grand Forks: University of North Dakota, Fisheries Research Unit.

Pearson, G. 1983. *A Review of the Impacts of the Garrison Diversion Unit on Fish and Wildlife Resources.* Prepared for the National Audubon Society, Jamestown, N.Dak.

Peet, S.E. and J.C. Day. 1980. The Long Lake diversion: an environmental evaluation. *Canadian Water Resources Journal* 5(3):34-48.

Petitjean, M.O.G. and Davies, B.R. 1986. Ecological impacts of interbasin water transfers: Some case studies, research requirements, and assessment procedures in southern Africa. *South African Journal of Science* 84:819-828.

Reynolds, P.J. 1986. Interbasin water transfers in Canada, past and future. In *Proceedings of IWRA Seminar on Interbasin Water Transfer,* 76-85. Beijing.

SCOPE (Scientific Committee on the Problems of the Environment). 1983. Newsletter, International Council of Scientific Unions.

Seagel, G.C. 1987. Pacific to Arctic transfer of water and biota, The McGregor diversion project in British Columbia, Canada. In *Proceedings of the Symposium on Interbasin Transfer of Water: Impacts and Research Needs for Canada*, 431-440. Saskatoon, Sask.: Canadian Water Resources Association.

Seagel, G.C. and M.G. Adam. 1986. A proposal for Pacific to Arctic water transfer, The McGregor Diversion Project in British Columbia, Canada. In *Proceedings of IWRA Seminar on Interbasin Water Transfer*, 351-360. Beijing.

Shafland, P.L., and W.M. Lewis. 1984. Terminology associated with introduced organisms. *Fisheries* 9(4):17-18.

Shiklomanov, I.A. 1985. Large-scale water transfer. In *Facets of Hydrology II*, ed. Roidda, J.C. 345-388. New York: John Wiley and Sons.

Shouquan, Z. 1983. Effect of diverting water from south to north on the ecosystem of the Huang-Huai-Hai Plain. In *Long-Distance Water Transfer: A Chinese Case Study and International Experiences*, ed. Biswas, A.K. 389-394. Dublin: Tycooly International Publishing Ltd.

Soucy, A. 1978. James Bay hydroelectric development. Evolution of environmental considerations. *Canadian Water Resources Journal* 3(4):54.

Templer, O.W. and L.V. Urban. 1995. The Canadian River project: A quarter century of interbasin transfer. *Proceedings of the American Water Resources Association Conference on Water Management in Urban Areas*, 93-102.

Timashev, I.E. 1982. Mezhbasseinovye perebroski rechnogo stoka v Kanade (Interbasin water transfers in Canada). *Izv. AN SSSR, ser. geogr.,* Moscow, (6):127-135.

Thomas, G.W. and T.W. Box. 1969. Social and ecological implications of water importation into arid lands. In *Arid Lands in Perspective*, 363-374. Tucson: University of Arizona Press.

Venugopalan, A.S. 1979. India's water wealth, its equitable distribution and utilization for optimum relief from floods and droughts. *Journal of the Institution of Engineers (India) Civ. Eng. Div.* 60(3):166-172.

Xuefang, Y. 1983. Possible effects of the proposed eastern transfer route on the fish stock of the principal water bodies along the course. In *Long-Distance Water Transfer: A Chinese Case Study and International Experiences*, ed. Biswas, A.K. 373-388. Dublin: Tycooly International Publishing Ltd.

Yuexian, X. and H. Jialian. 1983. Impact of water transfer on the natural environment. In *Long-Distance Water Transfer: A Chinese Case Study and International Experiences*, ed. Biswas, A.K. 159-168. Dublin: Tycooly International Publishing Ltd.

CHAPTER 4

Pathways for aquatic biota transfer between watersheds

Herbert R. Ludwig, Jr. and Jay A. Leitch

Executive Summary

Pathways for living aquatic organisms to move between adjacent watersheds are either naturally occurring or anthropogenic. Natural mechanisms include connections at times of high water, transport by animal (birds, terrestrial animals, and insects), and extreme meteorological events. Human induced transfer mechanisms were identified as either intentional or unintentional. Intentional mechanisms include authorized activities such as stocking fish for use as a sport fish, forage and baitfish, biological control species, and the reintroduction of endangered species. Unauthorized activities include release by aquarium hobbyists, aquaculture, biological control, and angler release. Unintentional activities include the release of aquatic species by anglers, transport of aquatic organisms on watercraft and angling equipment, road maintenance equipment, water projects, and activities associated with aquaculture. The annual probability of transport of aquatic organisms across watershed boundaries approaches 1.0 because of the large number of events. All of the pathways have some potential to serve as mechanisms to disperse aquatic species to adjacent watersheds. The identification of different mechanisms has two implications. First, concerns over the potential for the Garrison Diversion Unit to serve as a mechanism of aquatic introduction may be overstated. The activities of anglers, bait suppliers, aquaculturists, and even public fisheries managers may in the future contribute or may have already contributed to the movement of aquatic organisms between watersheds. Second, in light of this conclusion, the allocation of resources to reducing a single pathway, such as by filtering and treating Garrison Diversion water, without addressing others, may be ineffective in reducing the overall potential for inter-watershed transfer of aquatic species.

Introduction

The boundaries of major watersheds, continental divides, are biophysical barriers limiting the ranges of aquatic species. Neighboring watersheds often develop unique aquatic ecosystems (Clambey et al. 1983). Similarities between watersheds, such as common fish species, have been attributed to past connections during periods of glaciation and meltdown (Clambey et al. 1983). Present connections may exist between watersheds which may contribute to organism dispersal (movement away from point of origin) or intermingling. The purpose of this chapter is to identify and briefly describe possible pathways for aquatic biota transfer (the dispersal of living organisms across watershed boundaries) to occur (fig. 4.1). Before committing resources to control interbasin transfer of biota

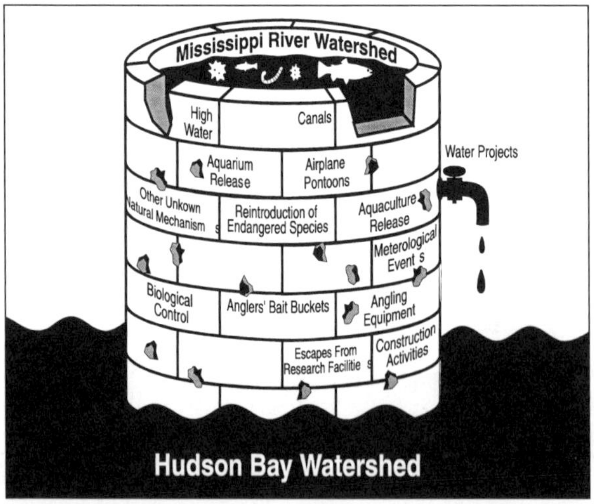

Figure 4.1 Potential pathways for living organisms to move between watersheds.

occurring through the Garrison Diversion Unit or any other source, it is important to identify all other possible biota transfer pathways. Efforts to reduce the probability of transfer via one mechanism and not others may do little to reduce the overall potential of interbasin transfer.

Pathways of biota transfer

Pathways of biota transfer are either naturally occurring or anthropogenic. Biophysical pathways from one watershed to another include animal transport, extraordinary meteorological events, intermittent connections at times of high water, and other unknown mechanisms. Human transport mechanisms can be either intentional or unintentional. Intentional introductions are usually attempts to "improve" existing conditions. They can be either unauthorized introductions by anglers and "amateur" zoologists or authorized stockings of species for sport, forage, bait, food, biological control, and reintroduction of an endangered species. Unintentional or accidental transfer mechanisms include activities associated with sportfishing, ballast from ships, construction activities, water projects, aquaculture, aquarium or ornamental species, and other mechanisms.

Barriers to aquatic introductions

Institutional and biophysical barriers restrict the transfer of aquatic biota between watersheds (fig. 4.2). Institutional barriers are rules and regulations, which serve as barriers to anthropocentric introductions; and many biophysical barriers exist which block the introduction of non-indigenous species.

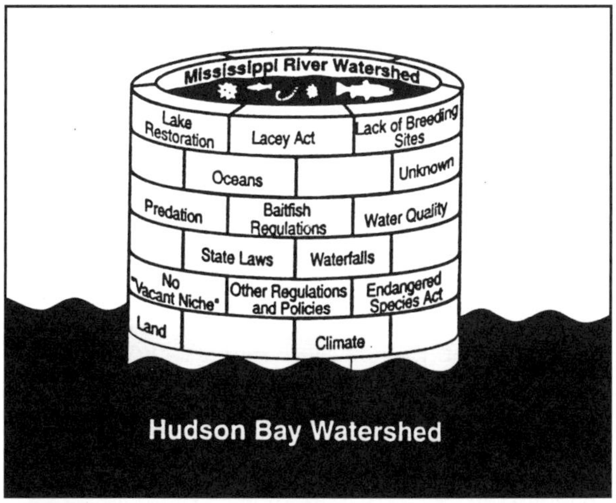

Figure 4.2 Barriers restricting the passage of living organisms between watersheds.

Institutional barriers to introduction

Institutional barriers block the transport or mitigate the effects of potentially invasive species. Several United States federal laws address the importation of exotic species. The first was the Lacey Act of 1900 which prohibits the import or export of any fish, wildlife, or plant taken, possessed, transported, or sold in violation of state, Indian tribe, foreign country, or treaty laws (Kurdila 1988). Several loopholes in the law motivated a series of amendments which substantially strengthened the Lacey Act.

Executive Order 11,987, signed by President Carter in 1977, restricted importation of exotic species to the United States. It was a noteworthy attempt at addressing the problems associated with exotic species, but was not implemented (Kurdila 1988). The Endangered Species Act of 1974 (U.S. Public Law 93-205) indirectly limits exotic species introduction. It prohibits the import, sale, and export of designated threatened and endangered species. Importers and exporters of fish, wildlife, and plants must receive permission from the Secretary of the Interior. Neither the Lacey Act nor the Endangered Species Act comprehensively addresses problems associated with the transport and subsequent introduction of exotic species. A major criticism of the Lacey Act is that it operates *ex post facto*. The penalties are enforced after the damage has occurred. The Endangered Species Act only addresses potential problems with the introduced species, or the presence of a threatened or endangered species in the receiving ecosystem.

States have developed regulations to address non-indigenous and exotic species importation and introductions. Most regulations have been general prohibitions against the importation and introduction of a specific species on a "dirty list." The "dirty list" approach, however, fails to address the importation of exotics which have yet to be identified as having the potential to harm state resources (Minnesota Interagency Exotic Species Task Force 1991).

Probably the strongest institutional barriers to movement and introduction of aquatic species are the rules and regulations that states place on the use, species, and import of baitfish. There are baitfish regulations in 54 of 57 United States states and Canadian provinces and territories (Litvak and

Mandrak 1993). Enforcement rigor and prevention of introductions vary from state to state (Office of Technology Assessment 1993). Restrictions on legal baitfish provide institutional barriers that limit the use of unwanted species of fish that may be introduced. For example, North Dakota has regulations that prohibit the use of white sucker *Catostomus commersoni* in many state waters (North Dakota Game and Fish Department 1994).

There are some practical institutional mechanisms which provide barriers between river basins by mitigating the presence of a transplanted species. For example, resource managers may have the option to entirely eliminate a transplanted aquatic species by killing off all of the fish in a body of water. This has been a common practice in North Dakota, where unwanted species populations equaled, or outnumbered, populations of desirable fish.

Even if all institutional barriers functioned ideally, they would not entirely limit the transport and introduction of aquatic species across watershed boundaries. Most state and provincial boundaries in the United States and Canada do not follow natural systems boundaries. Therefore, many species can be transported without crossing major political boundaries, but they may cross from one major watershed to another.

Biophysical barriers to introduction

Biophysical barriers to the introduction of aquatic species are a result of the physical environment or biological factors. Land is the greatest barrier to aquatic species movement between neighboring watersheds. With few exceptions, aquatic species cannot survive out of water to successfully cross long distances over land. Another limiting factor is climate. For example, it would be unlikely that reproducing populations of tropical species can become established in northern climates. It would be difficult for an African species of tilapia *Tilapia* Sp. to become established in North Dakota and survive the winter. Water quality may also serve as a barrier to the dispersal of species from one watershed to another. Oceans, waterfalls (Livingstone et al. 1982; Underhill 1957), and shallow waters (Neel 1985) may effectively block the passage of potential invaders between watersheds. Other factors, such as lack of suitable breeding sites and predators, may provide natural barriers to the success of invading species.

An aquatic species may be able to reach another watershed by anthropocentric or biophysical means, but may fail to survive in the receiving ecosystem because of a lack of a "vacant niche." There has been considerable debate regarding the role (and definition) of vacant niches in contributing to the establishment of introduced species. Identification of a vacant niche has occasionally been used as justification for the introduction of a non-indigenous species (Moyle et al. 1986; Simberloff 1981). As a barrier to introduction, the lack of "open niche" implies that an established species fulfills the ecological attributes of the potential invader (Levin 1989). Therefore, an aquatic species may simply be unable to compete for local resources, as they are already being used by other species.

Biophysical transfer mechanisms

Several biophysical mechanisms of transport across river basin boundaries can be identified. They may occur through geologic, climatic, or animal induced events.

Connections between watersheds during high water

About 9,000 years ago, at the end of the last ice age, some of the areas in North Dakota, Montana, Minnesota, and Manitoba presently draining into the Hudson Bay basin drained south into the Gulf of Mexico from glacial Lake Agassiz (fig. 4.3). As the glaciers receded, a northward drainage evolved, developing the Hudson Bay basin (Wills 1972) (fig. 2.5). Today, the landscape divisions between the Hudson Bay and Mississippi River basins are not clearly defined. Some of the United States Geological Survey topographic maps are ambiguous about the precise watershed boundaries because some areas are quite flat (Soil Conservation Service 1990). The maps characterize the watershed boundary as a "non-contributing" region ranging from less than 10 miles to nearly 50 miles wide.

For example, the basin boundary in Minnesota occurs in areas that are ". . . so level that many swamps and marshes drain with equal ease to either side" (Souris-Red-Rainy River Basin Commission 1972, Volume 1, p. 4). In 1938 and 1939, two

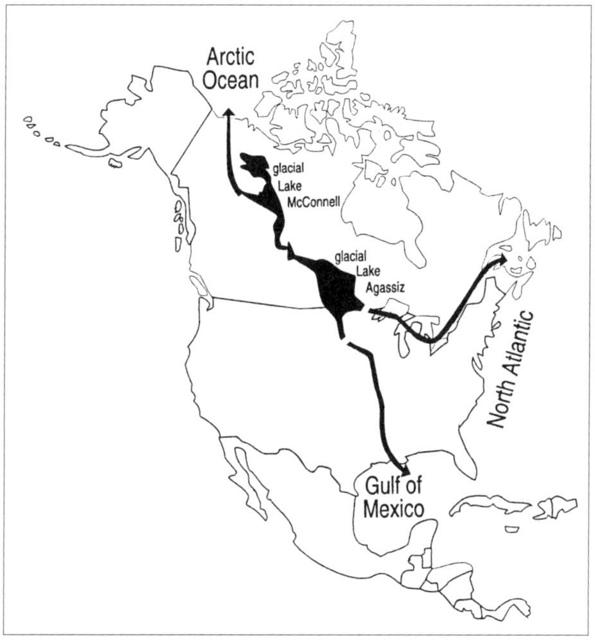

Figure 4.3 Drainage from Glacial-era Lake Agassiz.

walleye pike *Stizostedion vitreum*, which had been tagged the previous year in Cutfoot Sioux Lake in the Mississippi River basin, were captured in Bowstring Lake in northern Minnesota. "This is the only proof we have of the utilization of flooded low marshy areas in interbasin travels" (Underhill 1957, p.8). Otherwise, it appears that the opportunity for interbasin travel for mainstem biota is unlikely because of differences in water quality between the mainstem and headwaters (Neel 1985).

The close proximity of Lake Traverse, in the Hudson Bay basin, and Big Stone Lake, in the Mississippi River basin, makes a connection between the two a likely natural route for interbasin transfer and water-borne transportation. There is evidence that in 1820, settlers were able to pole and portage from Big Stone Lake to Lake Traverse. One optimistic businessman even attempted to bring a steamship into the Red River from the Little Mississippi River through this connection. The boat grounded before reaching Big Stone Lake and had to be abandoned (Remle n.d.).

The evidence is inconclusive regarding the efficacy of the connection between the two river basins before recent settlement. Control structures have now been constructed which keep the waters from intermingling except during extreme conditions. Those extraordinary conditions existed on July 25, 1993, when it was observed by Army Corps of Engineers personnel that water was flowing from the Little Minnesota River, through the Browns Valley box culvert, into Lake Traverse. This occurred within 36 hours, following a 10-inch rainfall near Sisseton, South Dakota. On August 6, 1993, water was observed running the opposite direction, from Lake Traverse into the Little Minnesota River basin. By August 13, no water could be observed flowing either way through the box culvert. Therefore, for several weeks, extreme conditions existed which facilitated the intermingling of the waters of the Mississippi River basin and the Hudson Bay basin (Bertschi 1994).

Animal transport

Biota transfer may occur when an animal carries an invasive species from one watershed to another. Birds, mammals, fish, insects, amphibians, and reptiles could each be vectors (Clambey et al. 1983).

Transport by birds

The watershed boundary in North Dakota and Minnesota is part of two major North American waterfowl flyways: the Mississippi and Central flyways (Smith et al. 1964) (fig. 4.4). With millions of waterfowl flying among the numerous lakes and potholes in close proximity to the basin boundary, the potential for birds to carry biota between lakes and, hence, to another watershed is clear. Three possible avian biota bearing transport mechanisms are (1) external adhesion, (2) regurgitation, and (3) within digestive tracts. External adhesion occurs when organic material becomes attached to the bird and the bird flies to a body of water where the attached biota is released. Birds have been observed to carry algae, crustaceans, aquatic plants, protozoans, and snails (Clambey et al. 1983 [who cites 15 other authors]). The occasional observations are not enough to convincingly establish that birds contribute to the dispersal of larger aquatic organisms such as fish through external adhesion.

There is speculation regarding the role of regurgitation of an aquatic species when an organism captured by a bird is released in another body of water (Clambey et al. 1983; Livingstone et al. 1982). This aspect of fish dispersal generally follows that a piscivorous (fish-eating) bird, such as a cormorant *Phalacrocorax* Sp., pelican *Pelicanus* Sp., or gull *Larinae* Sp., captures fish in one body of water, flies for some distance, and then releases the fish

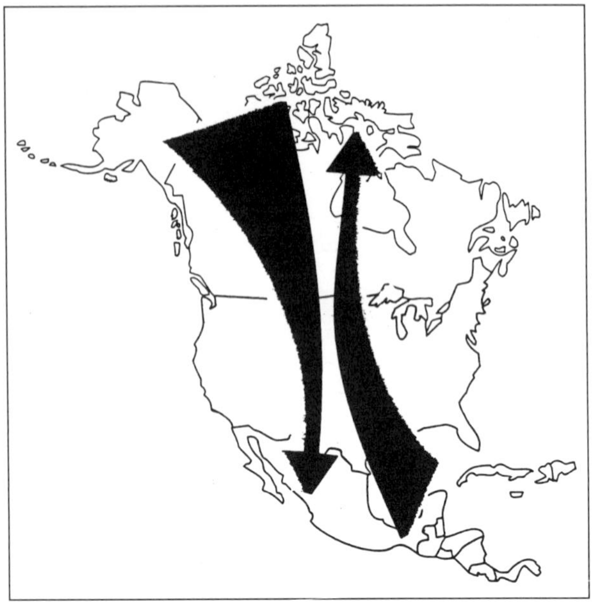

Figure 4.4 North American Waterfowl Flyways

into another body of water while regurgitating it for their young. Though regurgitation or release is a popular argument for the potential role of birds in transporting aquatic organisms, it has yet to be substantiated. There are several documented instances where nearby, and even connected, bodies of water (but broken by a waterfall) harbored numerous avian species, but contained discrete (different) populations of fish (Livingstone et al. 1982; Rickard et al. 1981; Balon 1974). Discrete populations, in such instances, suggest that even in cases of direct proximity, avian transport is not occurring.

The passage of undigested propagules through the digestive tract of a bird is a documented mechanism for the dispersal of plants, molds, and invertebrates. Studies of several species of birds and types of seeds have concluded that birds are viable mechanisms for the dispersal of propagules (Masaki et al. 1994; Herrera et al. 1994, Lambert and Marshall 1991). Research into the digestive systems of birds has supported the conclusion that birds retain seeds in their intestinal tract long enough for seeds to be transported several thousand miles (Clench and Mathias 1992, Proctor 1968). The cysts of an invertebrate (brine shrimp *Artemia salina*) ingested by mallard ducks *Anas platyrhynchos* were also shown to be able to remain viable (Malone 1965). Ground-feeding migratory song birds were observed to contribute to the intercontinental dispersal of spores, amoebae, macrocysts, and three propagules of slime molds (Suthers 1985).

Transport by terrestrial animals

Other animals may contribute to the dispersal of aquatic species. Several suggestions citing animals as possible mechanisms for aquatic species dispersal arise from the discovery of biota in the fur, intestines, or feces of animals. Beaver (Castoridae), muskrat *Ondatra zibethica* (Peck 1975), and crayfish (Decapoda) (Moore and Faust 1972) have been cited as possible mechanisms for species dispersal (Clambey et al. 1983). These animals have been observed to harbor crustaceans. Raccoons *Procyon* Sp. may also play an important role in the dispersal of aquatic organisms. A detailed ecological study on the overland colonization of small aquatic organisms such as paramecium and algae concluded that, "Raccoons are very effective in carrying many organisms simultaneously" (Maguire 1963, p. 184). The author further speculated that a raccoon could probably carry non-encysted algae, protozoa, etc., for many miles during a rainy night, and such an animal would visit small bodies of water seldom, if ever, visited by waterfowl (Maguire 1963).

Animals can also carry disease pathogens and parasites. The discovery of *Yersinia ruckeri*, a parasite which causes enteric redmouth in salmonids, in the lower intestine of a healthy muskrat has inspired some speculation regarding its possible role as a dispersal mechanism (Clambey et al. 1983). Though these are interesting observations, it is unclear how long a virus infecting poikilothermic (a cold blooded organism) host survives in a homiothermic animal. Reptiles and amphibians may also transport biota in a similar fashion. Turtles (Testudinata), snakes (Serpentes), salamanders (Caudata), frogs (Anura), and others may all contribute to the dispersal of small organisms.

Transport by insects

Maguire (1963) observed that captured dragonflies (*Dythemis, Plathemis, Libellula, Tramea,* and *Gomoides*), wasps (*Polistes, Pepsis*), a bee (Apoidea), and a cicada (Cicadidae) carried many species of aquatic organisms such as unicellular algae (12 [different kinds]), filamentous algae (8), Colpoda (8), other Ciliates (5), illoricate rotifer (5), and a small motile organism. Since dragonflies and butterflies can travel hundreds of miles, they are the most likely insects to function as long-distance dispersal agents for aquatic organisms (Maguire 1963).

Extreme meteorological events

Extreme meteorological events that transport organisms have been documented, but are not well understood due to their rarity (Maguire 1963). There are many anecdotal accounts of terrestrial "rains of fishes" (Gudger 1921) following a tornado or hurricane which began over water and expired over land. After researching and documenting 44 accounts of "rains of fishes," Gudger (1921, p. 619) concluded, ". . . waterspouts . . . passing over shallow water, would certainly pick up the small fishes swimming therein and, drawing them up into the clouds, would carry them over the country to drop them some distance away." Bajkov (1949) described an incident where seven species of fish fell from the sky to the main street of Marksvill, Louisiana. The incident was witnessed by bankers, merchants, and others who Bajkov describes as credible witnesses. Therefore, Bajkov concluded (p. 402) that "there is no reason for anyone to devaluate the scientific evidence. . ." that heavy wind can carry large objects, including fish.

Fish were found to be alive in many of these cases. Some were found frozen, others appear to have perished upon impact, and others were found already decomposing among live fish. One report in Gudger (1921) said that one fish among many others was fortunate enough to fall into a pool of water in a woodpile and was found swimming.

Other biophysical transfer mechanisms

High water connections, avian transport, animal transport, and extreme meteorological events are not the only possible biophysical introduction mechanisms. Possibilities for dispersal and introduction may arise from unforeseen circumstances and events which have not occurred, or been observed, during "modern" scientific observation. Volcanic eruptions (Gudger 1921), meteor impacts, earthquakes, extraordinary species behavior, or other biophysical phenomena, though highly improbable in this region and time and history, may contribute to the dispersal of species.

Anthropogenic transfer mechanisms

The introduction of non-indigenous species by humans has "drastically altered the distribution and abundance of living things and the structure and function of ecosystems on a truly global scale" (Brown 1989, p. 85). About 20 percent (151) of the 743 native fishes known to inhabit Canada and the continental United States have expanded their ranges through human activity (Courtenay and Taylor 1986). Anthropogenic introductions are either intentional or accidental. One important aspect of introductions of aquatic species is the unwanted species may "hitchhike" along as a parasite or in the water as a non-target species. A non-target species may accompany, and may be found within or on, a target species (an organism being intentionally transported) as parasites (Moyle and Leidy 1992), viruses (Farley 1992: Lightner et al. 1992: Gangzhorn et al. 1992: Hoffman and Shubert 1984: Shotts and Gratzek 1984) and even in the transport media (Carlton 1992, Shotts and Gratzek 1984).

Intentional introductions

Motivations for intentional introductions must be made explicit if policies regarding their control are to be implemented. Three agents can be identified that relate to three modes of intentional introduction. First, an introduction is made by a public agency without considering all possible consequences (authorized stocking activities). Second, an introduction is made by an individual who does not consider, or ignores, the possible consequences (unauthorized stocking activities). Third, the moral and ethical beliefs of individuals may override good ecological practice (such as the release of an aquarium fish or other captive organism).

Authorized introduction activities

Intentional aquatic introductions can be undertaken by a governing agency for a variety of reasons. Introduction of an aquatic species can serve, though not exclusively, as sportfish, forage or baitfish, food species, biological control agent, or progenitor for the reintroduction of an endangered species. Few introductions can be described as unequivocally successful, since it is difficult to foresee all possible consequences. One of the most descriptive references critical of intentional introductions of aquatic species is referred to as the "Frankenstein Effect" (Moyle et al. 1986, p. 415). This illustration borrows the theme from the 1818 Mary Shelley novel, *Frankenstein: or the Modern Prometheus* (Shelley 1831), where the scientist, Frankenstein, attempted to create an improved human being. This effort was soon discovered to have created more problems than it had solved. The consequences and moral dilemma that Frankenstein encountered parallel those of modern fisheries managers, whose decisions "are often made to improve local fisheries without consideration of their effects on native fish communities" (Moyle et al. 1986, p. 415). Examples of this effect are prevalent in the literature. Few exotics, namely brown trout *Salmo trutta*, bairdiella *Bairdiella icistia*, and orangemouth corvina *Cynoscion xanthulus,* have been considered beneficial (Welcomme 1984, 1981). Listings of introduced and exotic fish species can be found in the literature (OTA 1993; Courtenay and Kohler 1986; Courtenay and Taylor 1984, Welcomme 1984).

Sportfish

The introduction of non-indigenous fish for sportfish purposes has been the most common type of introduction (Courtenay 1993, Courtenay and Taylor 1984). The goal of legal sportfish introductions is to enhance angler opportunities, both for food and sport. Angling is an important recreational economic activity with over $US24 billion spent yearly (U.S. Department of the Interior 1993). Therefore, the motivation for fisheries managers to maintain productive fisheries is high. Often, this means that managers seek out fish that anglers find desirable and that also will be able to thrive in local conditions. Introduced sportfish include walleye *Stizostedion vitreum*, crappie *Pomoxis* Sp., yellow perch *Perca flavescens*, northern pike *Esox lucius*, and species of the salmon (Salmonidae) family.

Forage and baitfish

The motivation for public and private fisheries managers to introduce forage fish is to maintain productive sport fisheries. Forage fish serve as food for sportfish. Species introduced as forage include rainbow smelt *Osmerus mordax*, golden shiner *Notemigonus crysoleucas*, and fathead minnow *Pimephales promelas* (Higginbotham 1988).

Biological control

Introducing predators, pathogens, and competitors as biological control agents can serve to control unwanted organisms. The technique gained popularity in the United States in the late 1960s and early 1970s as an alternative to chemical treatments (Shireman 1984). Grass carp *Ctenopharyngodon idella* and various species of tilapia *Tilapia* Sp., for example, have been used to control aquatic plants (Shelton and Smitherman 1984). Experiments with biological control agents have been conducted legally and illegally and have resulted in establishment of unwanted and destructive populations (Courtenay 1993).

Aquatic biological agents have also been used to control insects that use aquatic environments for part of their life cycles. The most widely used fish, mosquitofish *Gambusia affinis*, has become established, often to the detriment of native species (Courtenay 1993). As a result, the use of indigenous fishes for mosquito *Culex* Sp. control is becoming more popular (Welcomme 1992). Microbial pathogens (viruses, protists, fungi, and bacteria) have also been used to control insects such as mosquitoes and black flies. Apparently, these agents have shown little effect on non-target organisms (Walton and Mulla 1992).

Reintroduction of endangered species

Occasionally aquatic species are introduced outside their native ranges for preservation. These kind of introductions are usually well planned and reviewed; and, therefore, few problems are encountered (Courtenay 1993). Two nonaquatic species identified as endangered are the black-footed ferret *Mustela nigripes* and the California condor *Gymnogyps californianus*. The black-footed ferret, having once lived successfully among vast colonies of prairie dogs in the Great Plains, is one of the most endangered mammals in North America. They fed on small reptiles, birds, and rodents—principally, the prairie dogs themselves. As farms, ranches, and other developments began to displace many of the prairie dog "towns," the black-footed ferret lost both its home and principal food source. Today, there are only a few tiny populations left and these were established with animals taken into and bred in captivity (Bosson 1992).

With a wingspread of eight to nine feet, and weighing twenty to thirty pounds, California condors have been on the verge of extinction for over ninety years. In 1985, only nine birds remained in the wild. These birds were brought into captivity, the last wild condor captured in 1987. Captive breeding has increased the condor population to 120. The first condors were released in the coastal mountains of central California in 1992, and there have been six other California releases since then. In December 1996, seven young condors were released in Arizona, marking the first time the birds had flown in Arizona since 1924. The government hopes to establish wild populations in California and Arizona, with a population of 150 birds at each site. Arizona may be the safer of the two release states as some of the previously released condors have died after coming into contact with power lines, antifreeze, and other manmade hazards (Burnham 1997). The California condor once ranged all along the Pacific Coast of North America and eastward to Texas. Never common, reproduction being extremely slow since they lay a single egg only every two years and mating does not occur until the young bird is five or six years old, the species is difficult to maintain. Parent condors will abandon an egg if people come too near the nesting sight. Egg collectors initially contributed to their endangered status, as well as careless hunting and poison bait often meant for other animals. Today, two nesting areas in California — Sisquoc and Sespe Condor Sanctuaries — are closed to the public (May 1972).

Unauthorized introduction activities

Individuals may conduct unauthorized fish introductions. Instances of these introductions are largely undocumented and are mostly anecdotal. These introductions may be motivated by good intentions, but are not based on sound management. Unauthorized introductions include those made by aquarium hobbyists, aquaculturists, and anglers.

Aquarium hobbyist releases

Aquarium hobbyists and ornamental fish farmers may release fish into the environment, often to the detriment of native and sometimes endangered species (Courtenay and Taylor 1986). About 46 of the 70 exotic fish established in the United States are ". . . known or believed to have escaped from aquarium fish culture facilities or were introduced by hobbyists" (Courtenay 1993, p. 49). The motives behind the deliberate release of an aquarium species are unclear. The most likely explanation may be that the person(s) releasing an aquarium species may have sentimental motivations regarding the preservation of the life of a pet (Courtenay 1993).

Introduced aquarium fishes become established mostly in subtropical areas, such as southern Florida, but also in the drainage canals in metropolitan areas, in remote desert ponds and lakes, and in thermal springs in northwestern states (Welcomme 1992). None of these introductions have had beneficial consequences. Some of the more notable examples of aquarium fish releases include snakeheads (Channa) in Maine and Rhode Island; piranhas *Serrasalmus* Sp. found in Florida, Michigan, and Ohio; and pacus (Colossoma) in many states (Courtenay 1993).

Aquaculture

Aquaculture, the production of fish and other seafood in controlled environments, has the greatest potential to introduce invasive fishes into North American waters if it remains unregulated (Courtenay 1993; Courtenay and Williams 1992). This seems likely because of three factors. First, the aquaculture industry is expanding. As U.S. consumers are becoming more health conscious, they are purchasing more fish as a low-calorie and low-fat source of protein. Second, a void is predicted in fish supply as wild-caught fish become relatively scarcer. This void may be filled by aquaculture (Hushank 1993, OTA 1993). Third, interest in finding novel and more productive species will continue to motivate aquaculturists to experiment with species (Courtenay 1993, OTA 1993).

An intentional aquaculture introduction could occur when stocking activities are conducted in open waters with non-native fish. Since aquaculture is a business, it is not rational for aquaculturists to "lose" any of their stock to the wild (Courtenay and Williams 1992). Thus, losses, or escapes, of fish from the controlled environment of aquaculture facilities are unintentional.

Non-indigenous and exotic species have been introduced into the wild to provide a human food source. Some introductions occurred at the frontiers of settlement of the United States. For example, exotic carp were introduced into the United States (Courtenay and Kohler 1986). Rivers were stocked by railway employees who dumped fish from railway bridges into bodies of water (Moyle 1986). Railway executives could then cite plentiful fish as an amenity when enticing immigrants to settle along railroad routes.

Biological control

Fish farmers have been known to establish fish populations where they later become "biological pollution" (Courtenay 1993). For example, establishment of grass carp in the U.S. has been viewed as detrimental. Egg thermal stocking techniques are used to sterilize fish to ensure that they do not become established in the wild (Shireman 1984). The search for new aquaculture species and for biological control agents is likely to continue (OTA 1993, Courtenay and Williams 1992). This search, without adequate precautions, could lead to establishment of undesirable aquatic species.

Angler release

Motivations behind the release of baitfish by anglers vary. It deserves mention as an unauthorized mechanism of introduction. Anglers may release extra baitfish, thinking that the fish will serve the ecosystem beneficially as forage (Litvak and Mandrak 1993). Anglers may also release fish to avoid killing them.

Accidental introductions

Accidental introductions arise when a species is found in waters where it is not indigenous as a result of unintended actions by humans. Humans may directly transport the organism, or they can provide the mechanism which enables organisms to enter a different body of water. This includes non-target species that accompany intentional introductions.

Since the effects of an introduction are usually only known after the fact, it is often difficult to determine the precise mechanism that led to its introduction. Consequently, the identification of the mechanism of introduction is based on a most likely mechanism with reference to other possible mechanisms (Mills et al. 1993). For example, Yount (1991) identified several ways that the zebra mussel could spread through human influence (recreational boating and angling, commercial vessels, man-made canals/irrigation ditches, aircraft pontoons, research equipment, and litter or garbage) and through biophysical mechanisms (water currents, external adhesion on waterfowl, and insects).

Release of aquatic species by anglers

Anglers are often attributed as likely mediums for unwanted introductions. Most evidence identifying anglers as mechanisms for the introduction of aquatic fishes is based on expert opinion (Carlton 1992, Courtenay 1993, Courtenay and Williams 1992, Courtenay and Kohler 1986, Courtenay and Taylor 1986, Mills et al. 1993, Moyle 1973, Welcomme 1992). Anglers have also been identified as likely mechanisms for the transport of non-target organisms (Carlton 1992, Yount 1991) and the transfer of species across river basin boundaries, or bait bucket transfer. Interest in the role of anglers as a mechanism for the introduction of non-indigenous species has lately generated two studies assessing the role of anglers regarding the transport of biota (Litvak and Mandrak 1993, Ludwig 1995).

Litvak and Mandrak (1993) studied the ecological impacts of the bait industry on donor and receiving ecosystems. They concluded that 18 of the 28 fish species found in bait dealers' tanks in Toronto, Canada, had the potential to be released outside their original ranges. Litvak and Mandrak's conclusion that anglers were responsible for range expansion was based on (1) the presence of separate populations of fish in Ontario and (2) 41 percent of anglers interviewed stating that they released unused baitfish.

Ludwig (1995) estimated the potential for bait bucket transfer to occur between the Mississippi River and Hudson Bay basins in North Dakota and Minnesota. By collecting data about the bait industry and anglers, and utilizing the conclusions of Litvak and Mandrak (1993), he showed the probability of bait bucket transfer between the two basins over a year's time to be close to one. The probability that anglers' bait buckets contain one of the species identified as a species of concern by the International Joint Commission (1990, 1976) was not investigated.

Restoration practices of the North Dakota Game and Fish Department also support the notion that bait bucket transfer is occurring. Restoration of angling waters usually follows when a body of water becomes "contaminated" by unwanted fishes. These unwanted fishes eventually crowd out the desirable fishes. For example, in North Dakota, one of the "undesirable fish," white sucker *Catostomus commersoni*, is a legal baitfish only in the Missouri and Red River systems. But it occasionally appears in other bodies of water, which had previously been restored, in sufficient numbers to warrant another restoration. Restoration usually involves chemically treating a body of water to kill all fish species. Following treatment, the body of water is stocked with desirable and compatible species. Anglers releasing baitfish are believed to be the source of reintroduction of the white sucker into these waters (Steinwand 1994). Non-baitfish such as bullhead *Ictalarus* Sp. and perch *Perca flavescens* may also be released by anglers if found within the bait bucket.

Introductions through watercraft and angling equipment

Watercraft and angling equipment may be mechanisms for transport of aquatic species. Aquatic species may be transported as part of the interior of watercraft, or may adhere externally to the watercraft or associated equipment (Gunderson 1995).

The release of ballast containing aquatic species from transoceanic ships is a documented mechanism for the introduction of exotic species (Garton et al. 1993; Mills et al. 1993; Carlton 1992). The introduction of the zebra mussel from Europe into the Great Lakes is attributed to the release of freshwater ballast (Yount 1991). Introductions from the ballast water from ships are not limited to mollusks, but may also include crustaceans (spiny water flea *Bythotrephes cederstroemi)* and fish (river ruffe *Gymnocephalus cernuus*, tubenose goby *Proterorhinus marmoratus*, and round goby *Neogobius melanostomus*).

Another source of transport of aquatic biota in watercraft may be anglers' live wells. Live wells maintain baitfish in a lively condition while in transit over land (or across a body of water) on a fishing vessel (either recreational or commercial). They may have refrigeration and aeration equipment to recirculate fresh water from the lake or river. In this way, target and non-target organisms may be released or escape from live wells far from their original sources (Carlton 1992).

Aquatic organisms may also be transported over land and across large bodies of water on angling equipment. One of the most commonly cited methods for aquatic plants to be transported between bodies of water is through the adhesion to boats and trailers. The nature of freshwater angling in the United States and Canada encourages the movement of anglers between bodies of water. Anglers in North Dakota travel an average of 33 miles to an angling destination (Baltezore and Leitch 1992). The spread of Eurasian watermilfoil *Myriophyllum spicatum* in Midwest lakes is attributed to itinerant boat trailers (OTA 1993). Mollusks, such as the zebra mussel, can also adhere to the exterior of boats traveling both in the water and being transported over land. This may have contributed to their rapid dispersal in the Great Lakes (Yount 1991) and its potential dispersal into other bodies of water such as in California (U.S. Water News 1995). Seaplane pontoons may also lead to zebra mussel introduction in inland water bodies far from its source in the Great Lakes (Yount 1991).

Road maintenance

Maintenance of roads may facilitate the transport of aquatic organisms through the use of water to control dust. It is a common practice in rural areas to control dust by spraying water taken from a convenient surface source onto unpaved roads. It is possible that such spray water may enter other bodies of water and hence introduce accompanying aquatic species.

Water projects

There are many different ways that interbasin water transfers can occur with water projects. Source waters can be surface water, such as a river or lake, or a managed supply source, such as sewage or municipal water supply. Surface-source interbasin transfers usually occur on a large scale for navigation, irrigation, or municipal water supply, whereas managed sources may transfer water only across a town that is located on the basin boundary. Some of the more notable examples of large scale interbasin transfer for irrigation and municipal purposes are the transfers occurring from the Colorado River basin to farms and metropolitan areas in southern California (Committee on Western Water Management 1992). Minnesota has catalogued seven interbasin transfers that occur from the Hudson Bay basin into the Mississippi River basin. One interbasin transfer for sewage, in the town of New York Mills, Minnesota, occurs from the Mississippi River basin into the Hudson Bay basin (Trotta 1988).

Bodies of water that have been altered by human activity are vulnerable to the establishment of introduced species (Brown 1989; Moyle and Leidy 1992; Moyle et al. 1986). Therefore, interbasin transfers are particularly menacing because of the double effect of altering system ecologies, which facilitates the introduction of a non-indigenous species while simultaneously providing a conduit for its movement.

Releases from aquaculture

Much scrutiny has lately been given to the activities of the aquaculture industry and its role in the transfer of non-indigenous species. Nearly all exotic aquaculture species have escaped into open waters (Courtenay and Williams 1992). Therefore, aquaculture species should be considered a potential addition to the fauna of nearby waters (Courtenay and Williams 1992).

Food fish, baitfish, sportfish, biological control fish, and aquarium fish can be raised in an aquaculture facility; and the escape of aquarium species from aquaculture facilities has been well documented. There are several ways that fish can accidentally escape from an aquaculture facility. They can escape through unscreened effluent pipes in both indoor and outdoor facilities, when floods inundate containment facilities, and with the release of excess stock into adjacent waters (Courtenay 1993). Escapes from either aquaculture facilities for aquarium fish and through intentional releases by hobbyists comprise 65 percent of the exotic fish species established in the United States (Courtenay 1993). Imported aquarium species may also contain non-target species. Tests undertaken on imported aquarium shipments indicated that pathogens may be included (OTA 1993).

The aquaculture industry has been responsible for transport of non-target species. Non-target species may be any other species which was overlooked during collection of the target species, smaller organisms such as parasites that would escape unnoticed without rigorous inspection (OTA 1993), fish pathogens (Gangzhorn et al. 1992), invertebrate pathogens (Lightner et al. 1992; Farley 1992) algae (Carlton 1992), and other living organisms. A non-target species may accompany a target species as an associated species, as biota in/on a target or non-target species, and as biota in/on transport media (Carlton 1992).

Discussion

The movement of aquatic organisms across watershed boundaries is inevitable given enough time. Several mechanisms of transfer were identified and described. Of these, five were biophysical mechanisms of introduction, and the rest were anthropogenic.

The Mississippi River and Hudson Bay watersheds appear to be particularly vulnerable to biophysical mechanisms of introduction. The vulnerability of the region can be attributed to the wide expanse of relatively flat terrain that divides the two regions. High water connections, therefore, have a potential to contribute to the dispersal of aquatic species. The capture of Mississippi River watershed tagged walleye in the Hudson Bay watershed, and the recent flows between Lake Traverse and Big Stone Lake, illustrates the potential role high water has in species dispersal.

Any animal that visits two different bodies of water has the potential to transport and introduce aquatic organisms into new environments. Though transport of aquatic organisms by animals is pervasive, the size of the transported organisms appears to be a limiting factor. The presence of a major avian flyway crossing the watershed boundaries enhances the potential for birds to act as transfer mechanisms between the Mississippi River and Hudson Bay watersheds.. Flying insects may also harbor aquatic biota. Similarly, terrestrial animals are possible mechanisms for the introduction of aquatic species.

Extreme meteorological events may result in the transfer of aquatic organisms from one watershed to another. Tornadoes and high winds are common in North Dakota and Minnesota. Since rural areas are remote and sparsely populated, "rains of fishes" may occur without observation.

National statistics of aquatic introductions reveal that aquaculture-related activities may result in the greatest proportion of aquatic introductions. Most aquaculture introductions have been the result of escapes from facilities that propagate tropical fish for ornamental purposes. Ornamental fish introductions, though common in tropical areas, are unlikely to occur - or at least survive - in the Hudson Bay watershed and the northern extremes of the Mississippi River watershed. The potential for transfer of a non-indigenous target species from one watershed to another by the aquaculture industry appears low. Reports indicate that many of the fish raised in the area already exist in both the Hudson Bay watershed and the Mississippi River watershed (Minnesota Department of Agriculture 1993). An exception, tilapia, an exotic fish cultured in North Dakota, is unlikely to pose a threat to endemic fisheries since it is confined to warmer effluent waters of a power plant.

The dissemination of non-target species associated with the transport of an aquaculture species is possible for two reasons. As the industry expands, the exchange of products also expands. Greater volumes of goods are exchanged with increasing frequency which may increase the chances of inclusion of non-target species. Also, transport regulations govern aquaculture within political boundaries. There are no regulations pertaining to the transport of aquatic species across watershed boundaries within political boundaries.

The transfer of target and non-target species across watershed boundaries by anglers is likely. Anglers are thought to be responsible for the release and subsequent establishment of many discrete populations of fish in many bodies of water. Often, these populations become established in bodies of water that have restricted the use of bait or the species of bait. The transport of a non-target fish species is also likely. Bait is not perfectly sorted by bait vendors, and the transported water may contain other kinds of biota, including other vertebrates, invertebrates, and plants.

The transfer of biota is likely as a result of the construction of water projects. Interbasin water transfers and canal construction have resulted in introductions. With the construction of canals and other transportation-related conduits, the movement of ships and associated ballast water also carries the possibility of biota transfer.

Angling equipment, boats, trailers, construction equipment, aircraft pontoons, and virtually any object moved from one body of water into another have potential to carry invasive biota. Though these activities may, arguably, have a low probability of transferring biota in a single event, their cumulative probability is high due to the large number of events.

Conclusions and implications

Aquatic organisms can cross river basin boundaries via several biophysical or anthropogenic pathways. Biophysical barriers can be breached during high water or during extreme meteorological events (e.g., tornado). Anthropogenic pathways across biophysical barriers include those that are accidental as well as those that are intentional. Human interference has hastened the range expansion of many species, which raises concerns among ecologists and others (Levin 1989).

For each pathway of biota transfer, there is some probability greater than zero that it will contribute to the range expansion of biota across basin boundaries. The geographic extent, species involved, and numerical risk/probability associated with each pathway has not been specifically addressed. Some pathways, such as anglers' bait buckets, are more obvious than others and have received frequent mention in the literature. Others, such as using surface waters for dust abatement during road construction, are less obvious.

The probability of bait bucket transfer was identified as approaching 1.0 when all the individual events over a year were aggregated. However, given the specific aquatic niches and life-cycle sensitivities of some of the potentially invasive species, the probability of introduction may be considerably lower for some species via some pathways than for others. This analysis implicitly assumed a random distribution of species across aquatic niches in the basin.

The effects of range expansions into adjoining watersheds have been largely negative, although positive effects can be found. There is general agreement that increased analyses need to be conducted to develop effective protocols regarding introductions (Courtenay and Taylor 1986). Careful monitoring, contingency plans for biological or physical containment, and plans to mitigate negative side effects are necessary prior to introductions (Levin 1989).

Efforts to prevent biota transfer via a single pathway need to be considered in light of the potential for transfer via other pathways. For example, assume that an invasive species has an equal probability of being transferred across a watershed boundary via any one of a dozen pathways. Whether that probability was high or low, it would be ineffective to block anything less than most or all of the pathways. Damage is caused as a result of a single successful breach. However, multiple breaches may accelerate the damage in time or across space.

These observations have important implications for the proposed Garrison Diversion Unit in North Dakota. Water projects, such as the Garrison Diversion Unit, are one of the pathways of aquatic biota transfer across basins and add to the overall probability of eventual biota transfer. Treatment of Garrison Diversion Unit water, however, may not be an effective or efficient solution to preventing all aquatic biota transfer since other pathways exist, such as bait buckets. Resources and regulations should be directed to where they have the greatest impact on meaningful reductions in probabilities.

The potential for biota transfer through the Garrison Diversion Unit may be overemphasized. The probability of introduction via bait buckets and the aquaculture industry, which draw and release water (and living species) from many sources, would be higher than the Garrison Diversion Unit which draws water from a single point source. If one of the species of concern were within the bait bucket or aquaculture shipment, and brought to different basins, then its potential introduction would be more widespread spatially throughout the basin. With more widespread potential for introduction, one would expect more incidences of introduction into areas where conditions may be more ideal for colonization.

Similarly, anglers release unused baitfish into lakes, streams, and rivers which are relatively more amenable to colonization than the proposed structure of the Garrison Diversion Unit. Fish destined for passage through the Garrison Diversion Unit, under current design proposals, will be impeded by control structures. This argument does imply that there is a zero probability that fish or pathogens will not pass through the Garrison Diversion Unit; but in relative terms, passage will most likely occur by angling and aquaculture activity. Bait fish providers and aquaculturalists target biota (living animals) when collecting their product, whereas, the Garrison Diversion Unit intake will have a series of measures to ensure that biota does not freely use the conduit.

Information about target bait and non-target species (such as accompanying pathogens) is not available for the region. More work could be done to specify the aquatic habitat niches inhabited by organisms of concern to more clearly identify the relationships between those organisms and the pathways they may take to breach watershed boundary barriers. With this information, specific probabilities and pathways could be assigned for each of the species of concern as part of a comprehensive risk analysis on the nine species of concern. Application of appropriate policy measures could then be instituted, based upon relative measures of risk.

References

Balon, E.K. 1974. Fishes from the edge of Victoria Falls, Africa: Demise of a physical barrier for downstream invasions. *Copeia* 1974:643-660.

Baltezore, J.F. and J.A. Leitch. 1992. *Characteristics, Expenditures, and Economic Impact of Resident and Nonresident Hunters and Anglers in North Dakota, 1990-91 Season.* Fargo: Department of Agricultural Economics, Experiment Station, North Dakota State University.

Bertschi, T. 1994. Letter dated January 12, 1994. Western Area Resource Manager, United States Army Corps of Engineers, Fargo, N.D.

Bosson, J. 1992. *Wild and Free: The Story of a Black-footed Ferret.* Norwalk, Conn.: Trudy Management Corporation.

Brown, J.H. 1989. Patterns, modes and extents of invasions by vertebrates. In *Biological Invasions: a Global Perspective*, ed. Drake, J.A., et al., 85-109. New York: John Wiley & Sons.

Burnham, B. 1997. Condor killed by eagle in northern Arizona. *The Brookings (South Dakota) Register.*

Carlton, J.T. 1992. Dispersal of living organisms into aquatic ecosystems as mediated by aquaculture and fisheries activities. In *Dispersal of Living Organisms into Aquatic Ecosystems*, eds. Rosenfield, A., and R. Mann, 13-46. College Park: Maryland Sea Grant.

Clambey, G.K., H.L. Holloway, Jr., J.B. Owen, and J.J. Peterka. 1983. *Potential Transfer of Aquatic Biota Between Drainage Systems Having No Natural Flow Connections.* Fargo, N.Dak.: Tri-College University Center for Environmental Studies.

Clench, M.H. and J.R. Mathias. 1992. Intestinal transit: How can it be delayed long enough for birds to act as long-distance dispersal agents? *The Auk* 109(4):933-936.

Committee on Western Water Management. 1992. *Water Transfers in the West: Efficiency, Equity, and the Environment.* Washington, D.C.: National Academy Press.

Courtenay, W.R., Jr. 1993. Biological pollution through fish introductions. In *Biological Pollution: The Control and Impact of Invasive Exotic Species*, ed. McKnight, B.N., 35-61. Indianapolis: Indiana Academy of Science.

Courtenay, W.R., Jr. and C.C. Kohler. 1986. Exotic fishers in North American fisheries management. In *Fish Culture in Fisheries Management*, ed. Stroud, R.H., 401-413. Bethesda, Md.: American Fisheries Society.

Courtenay, W.R., Jr., and J.N. Taylor. 1984. The exotic ichthyofauna of the contiguous United States with preliminary observations on intranational transplants. In *Documents Presented at the Symposium on Stock Enhancement in the Management of Freshwater Fisheries*, 2:466-487. Rome: FAO, European Inland Fisheries Advisory Commission (EIFAC).

Courtenay, W.R., Jr., and J.N. Taylor. 1986. Strategies for reducing risks from introductions of aquatic organisms: A philosophical perspective. *Fisheries* 11(2):30-33.

Courtenay, W.R., Jr. and J.D. Williams. 1992. Dispersal of exotic species from aquaculture sources, with emphasis on freshwater fishes. In *Dispersal of Living Organisms into Aquatic Ecosystems*, eds. Rosenfield, A., and R. Mann, 49-81. College Park: Maryland Sea Grant.

Eckstein, O. 1971. *Water Resource Development: the Economics of Project Evaluation.* Cambridge: Harvard University Press.

Farley, A.C. 1992. Mass mortalities and infectious lethal diseases in bivalve mollusks and associations with geographic transfers of populations. In *Dispersal of Living Organisms into Aquatic Ecosystems*, eds. Rosenfield, A., and R. Mann, 139-154. College Park: Maryland Sea Grant.

Gangzhorn, J., J.S. Rohovec, and J.L. Fryer. 1992. Dissemination of microbial pathogens through introductions and transfers of finfish. In *Dispersal of Living Organisms into Aquatic Ecosystems*, eds. Rosenfield, A., and R. Mann, 175-192. College Park: Maryland Sea Grant.

Garton, D.W., D.J. Berg, A.M. Stoeckmann, and W.R. Haag. 1993. Biology of recent invertebrate invading species in the Great Lakes: The spiny water flea, *Bythotrephes cederstroemi*; and the zebra mussel, *Dreissenna polymorpha.* In *Biological Pollution: The Control and Impact of Invasive Exotic Species*, ed. McKnight, B.N. Indianapolis: Indiana Academy of Science.

Gudger, E.W. 1921. Rains of fishes. *Natural History* 607-619.

Gunderson, J. 1995. Three-state exotic species boaters survey: What do boaters know and do they care? *Aquatic Nuisance Species Digest* 1(1):8-10.

Herrera, C.M., P.Jordano, L. Lopez-Soria, and J.A. Amat. 1994. Recruitment of mast-fruiting, bird-dispersed tree: Bridging frugivore activity and seedling establishment. *Ecological Monographs* 64(3):315-345.

Hoffman, G.L. and G. Schubert. 1984. Some parasites of exotic fishes. In *Distribution, Biology, and Management of Exotic Fishes*, eds. Courtenay, W.R., and J.R. Stouffer, 233-261. Baltimore: Johns Hopkins University Press.

Hushank, L.J. 1993. *North Central Regional Aquaculture Industry Situation and Outlook Report.* Ames, Iowa: North Central Regional Aquaculture Center.

International Joint Commission. 1976. International Garrison Diversion Study Board. Appendix C, Biology Report.

International Joint Commission. 1990. Joint Technical Committee Report: Garrison Diversion Unit.

Kurdila, J. 1988. The introduction of exotic species into the United States: There goes the neighborhood! *Boston College Environmental Affairs Law Review* 16(1):95-118.

Lambert, F.R. and A.G. Marshall. 1991. Keystone characteristics of bird-dispersed ficus in a Malaysian lowland rain forest. *The Journal of Ecology* 79(3):793-800.

Levin, S.A. 1989. Analysis of risk for invasions and control programs. In *Biological Invasions: a Global Perspective*, eds. Drake, J.A., and H.A. Mooney, 425-435. New York: John Wiley & Sons.

Lightner, D.V., R.M. Redman, T.A. Bell, and R.B. Thurman. 1992. Geographic dispersion of the viruses IHHN, MBV and HPV as a consequence of transfers and introductions of penaeid shrimp to new regions for aquaculture purposes. In *Dispersal of Living Organisms into Aquatic Ecosystems*, eds. Rosenfield, A. and R. Mann, 155-173. College Park: Maryland Sea Grant.

Litvak, M.K. and N.E. Mandrak. 1993. Ecology of freshwater bait fish use in Canada and the United States. *Fisheries* 18(12):6-12.

Livingstone, D.A., M. Rowland, and P.E. Bailey. 1982. On the size of African riverine fish faunas. *American Zoologist* 22:361-369.

Ludwig, H. Jr., 1995. *Bait Bucket Transfer Potential Between the Mississippi and Hudson Bay Watersheds.* M.S. thesis, North Dakota State University, Fargo.

Maguire, B. Jr.,1963. The passive dispersal of small aquatic organisms and their colonization of isolated bodies of water. *Ecological Monographs* 33(2):161-185.

Malone, C.R. 1965. Dispersal of plankton: Rate of food passage in mallard ducks. *Journal of Wildlife Management* 29(3):529-533.

Masaki, T., Y. Kominami, and T. Nakashizuka. 1994. Spatial and seasonal patterns of seed dissemination of *Cornus Controversa* in a temperate forest. *Ecology* 75(7):1903-1911.

May, J. 1972. *Giant Condor of California.* Mankato, Minn.: Creative Educational Society, Inc.

Mills, E.L., J.H. Leach, J.T. Carlton, and C.L. Secor. 1993. Exotic species in the Great Lakes: A history of biotic crises and anthropocentric introductions. *Journal of Great Lakes Research* 19(1):1-54.

Minnesota Department of Aquaculture. 1993. *Minnesota Aquaculture Report.* St. Paul: Minnesota Department of Agriculture.

Minnesota Interagency Exotic Species Task Force. 1991. *Report and Recommendations*, St. Paul: Minnesota Department of Natural Resources.

Moore, W.G. and B.F. Faust. 1972. Crayfish as possible agents of dissemination of fairy shrimp into temporary ponds. *Ecology* 53:314-316.

Moyle, P.B. 1973. Ecological segregation among three species of minnows (Cyprinidae) in a Minnesota lake. *Transactions of the American Fisheries Society* 102(4):794-805.

Moyle, P.B. 1986. Fish introductions into North America: Patterns and ecological impact. In *Ecology of Biological Invasions of North America and Hawaii*, eds. Mooney, H.A., and J.A. Drake, 27-43. New York: Springer Verlag.

Moyle, P.B. and R.A. Leidy. 1992. Loss of biodiversity in aquatic ecosystems: Evidence from fish faunas. In *Conservation Biology: The Theory and Practice of Nature Conservation, Preservation, and Management*, eds. Fiedler, P.L., and S.K. Jain, 127-169. New York: Routledge, Chapman & Hall Inc.

Moyle, P.B., H.W. Li, and B.A. Barton. 1986. The Frankenstein effect: Impact of introduced fishes on native fishes in North America. In *Fish Culture in Fisheries Management*, ed. Stroud, R.H., 415-426. Bethesda, Md.: American Fisheries Society.

Neel, J.K., Sr. 1985. *A Northern Prairie Stream.* Grand Forks: University of North Dakota Press.

North Dakota Game and Fish Department. 1994. *North Dakota Fishing Regulations.* Bismarck.

Office of Technology Assessment (OTA). 1993. *Harmful Non-indigenous Species in the United States.* Washington, D.C.: United States Congress, OTA-F-565. GPO.

Peck, S.B. 1975. Amphipod dispersal in the fur of aquatic mammals. *Canadian Field-Naturalist* 89:181-192.

Proctor, V.W. 1968. Long-distance dispersal of seeds by retention in digestive tract of birds. *Science* 160:321-322.

Remle, L. n.d. Unpublished document. Moorhead, Minn.: Heritage-Hjemkomst.

Rickard, W.H., R.E. Fitzner, and C.E. Cushing. 1981. Biological colonization of an industrial pond: Status after two decades. *Environmental Conservation* 8:241-247.

Shelley, M. 1831. *Frankenstein; or the Modern Prometheus*. New York: Oxford University Publishers.

Shelton, W.L. and R.O. Smitherman. 1984. Exotic fishes in warmwater aquaculture. In *Distribution, Biology, and Management of Exotic Fishes*, eds. Courtenay, W.R., Jr., and J.R. Stauffer, 262-301. Baltimore, Md.: Johns Hopkins University Press.

Shireman, J.V. 1984. Control of aquatic weeds with exotic fishes. In *Distribution, Biology, and Management of Exotic Fishes*, eds. Courtenay, W.R., Jr., and J.R. Stauffer, Jr., 302-312. Baltimore, Md.: Johns Hopkins University Press.

Shotts, E.B., Jr. and J.B. Gratzek. 1984. Bacteria, parasites, and viruses of aquarium fish and their shipping waters. In *Distribution, Biology, and Management of Exotic Fishes*, eds. Courtenay, W.R., Jr., and J.R. Stauffer, Jr., 215-232. Baltimore, Md.: Johns Hopkins University Press.

Simberloff, D. 1981. Community effects of introduced species. In *Biotic Crises in Ecological and Evolutionary Time*, ed. Nitecki, T.H., 53-81. New York: Academic Press.

Smith, A.G., J.H. Stoudt, and J.B. Gollop. 1964. Prairie potholes and marshes. In *Waterfowl Tomorrow*, ed. Linduska, J.P., 39-50. Washington, D.C.: The United States Department of the Interior, Bureau of Sport Fisheries and Wildlife, Fish and Wildlife Service, GPO.

Soil Conservation Service. 1990. *Hydrologic Unit Map 1990 State of North Dakota*. Denver, Colo.: 1006438-01, United States Department of Agriculture.

Souris-Red-Rainy River Basin Commission. 1972. *The Combined Report - Type I Framework Study*, Vol. 1, 216 pp.

Steinwand, T. 1994. Letter dated July 7. Chief, Fisheries Division, Bismarck: North Dakota Game and Fish Department.

Suthers, H.B. 1985. Ground-feeding migratory songbirds as cellular slime mold distribution vectors. *Oecologia* 65:526-530.

Trotta, L.C. 1988. *Inventory of Interbasin Water Transfers in Minnesota*. St. Paul, Minn.: United States Geological Survey.

Underhill, J.C. 1957. *The Distribution of Minnesota Minnows and Darters: Relation to Pleistocene Glaciation*. Minnesota Museum of Natural History, Occasional Paper Number 7. Minneapolis: The University of Minnesota Press.

U.S. Department of the Interior, Fish and Wildlife Service and the United States Department of Commerce, Bureau of the Census. 1993. *1991 National Survey of Fishing, Hunting, and Wildlife Associate Recreation*. Washington, D.C.: GPO.

U.S. Water News. 1995. Zebra mussels sighted in Calif. where they could wreak havoc. *U.S. Water News* October 12(4):4.

Walton, W.E. and M.S. Mulla. 1992. Impacts and fates of microbial pest-control agents in the aquatic environment. In *Dispersal of Living Organisms into Aquatic Ecosystems*, eds. Rosenfield, A., and R. Mann, 205-237. College Park: Maryland Sea Grant.

Welcomme, R.L. 1981. *Register of International Transfers of Inland Fish Species*. Rome: FAO.

Welcomme, R.L. 1984. International transfers of inland fish species. In *Distribution Biology and Management of Exotic Fishes*, eds. Courtenay, W.R., Jr., and J.R. Stauffer, Jr., 22-40. Baltimore, Md.: Johns Hopkins University Press.

Welcomme, R.L. 1992. A history of international introductions of inland aquatic species. In *Introductions and Transfers of Aquatic Species*, eds. Steinmetz, S.B., and W. Hershberger, 194:3-14. International Council for the Exploration of the Sea.

Wills, B.L. 1972. *North Dakota: The Prairie State*. Ann Arbor, Mich.: Edwards Brothers.

Yount, D.J. 1991. *Ecology and Management of the Zebra Mussel and Other Introduced Aquatic Nuisance Species*. Duluth, Minn.: U.S. Environmental Protection Agency, Environmental Research Laboratory.

CHAPTER 5
Consequences of non-indigenous species
Jay A. Leitch

Executive Summary

Non-indigenous species invasions can result in beneficial, benign, or harmful impacts to both ecologic and economic systems (U.S. Congress, Office of Technology Assessment, 1993). The impacts can be obvious, direct, and immediate, such as with the zebra mussel. Or, they may be subtle, indirect, and latent as with about 90 percent of non-indigenous species introductions. In fact, OTA claims that high negative impacts are infrequent (OTA 1993). Ecologic impacts may include changes in species composition and abundance or in biodiversity. Economic impacts include changes in commercial or leisure time activity, or human health effects.

Ecologic consequences

While most non-indigenous species introductions turn out to be benign or subtle, some result in serious, widespread, and long-term changes in biotic communities and community biodiversity. Elton (1958) and OTA (U.S .Congress, OTA 1993) provide a thorough discussion of the ecologic consequences of non-indigenous species invasions.

A species invasion is the movement of a living organism into an area where it was previously absent. Until recently, the range expansions of species in North Dakota, especially those of European origin, have been viewed as beneficial, both as enhancements of the environment and for economic reasons (Crosby 1986). However, this attitude has changed as invasions of non-indigenous species have occurred, often at the expense of natural systems.

River ruffe *Gymnocephalus cernuus* is an example of a Great Lakes introduction which has impacted indigenous fishery resources. Ruffe arrived in Lake Superior in the ballast water of ships originally from European ports. First discovered in Duluth-Superior Harbor in 1986 and Thunder Bay, Ontario, in 1991, ruffe have increased in abundance and expanded their range. The presence of ruffe has been shown to decrease the abundance of indigenous fish, such as yellow perch *Perca flavescens*, trout perch *Percopsis omiscomaycus*, spottail shiner *Notropis spilopterus*, and emerald shiner *Notropis atherinoides*, by about 75 percent. Ruffe have long dorsal and anal spines making them a poor food source for larger fish. Subsequently, as they displace endemic forage species, there is a negative impact on the abundance of larger predatory fish such as walleye and trout. Their range is also expanding along the coast of Lake Superior.

Species considered invasive may not always be exotic or non-indigenous. For example, there are several angling waters in North Dakota managed for a particular type of fishery, such as trout or walleye. Several of these bodies of water, particularly those managed for trout, had been previously treated with rotenone. The chemical treatment, when applied properly on small lakes, destroys all living fish. Afterwards, desirable species are reintroduced. If undesirable native fish such as white sucker *Catostomus commersoni*, yellow perch, or bullhead *Ictalurus* Sp. become introduced (possibly through angler release from a bait bucket), they may grow and eventually crowd out more desirable species. Subsequently, the costly treatment process may have to be repeated.

Introduction of brown trout *Salmo trutta* into many water bodies in North America is regarded as one of the more successful and beneficial exotic fish introductions because of its popularity with anglers. But, the brown trout's establishment poses a threat to native fish populations, and many fisheries agencies have programs for its control and eradication. Though it is possible that introductions may be beneficial, whether intentional or unintentional, they pose a potential ecological and economic hazard. This hazard arises from the considerable amount of uncertainty regarding the ultimate effect of an introduction (Levin 1989, Moyle et al. 1987; Courtenay and Taylor 1984). Ecologically, introduced species

- may not survive;
- may not breed in natural conditions and must be maintained artificially by continuous introductions, such as triploid grass carp *Ctenopharyngodon idella*;
- may become locally established in an unusual habitat, such as thermally polluted areas or cold, high-altitude waters in tropical areas;
- may increase rapidly in abundance, then decline in abundance and disappear;
- may maintain a widespread or isolated population which has negligible environmental or economic effects; or have an impact on the ecological and economic elements of the receiving waters (Welcomme 1992).

Since there is considerable uncertainty regarding most species introductions *ex ante*, fisheries biologists call for greater study of the potential impacts of an introduction (Courtenay 1993, Welcomme 1992).

Time is an important consideration when assessing the impact of an introduction. For example, when the common carp *Cyprinus carpio* was introduced in the early 1800s, it was hailed as a "patriotic act" (Mills et al. 1993). Now, many fisheries managers decry the carp for its propensity to uproot plants, affect water quality, and consume eggs of native or game fish (Courtenay 1993).

There may also be a shortage of information surrounding the dispersal characteristics of an invading species. This may be particularly important if it can be determined that the invading species may overcome and occupy the ecological niche of a beneficial endemic species. For example, McCulloch and Stewart (1992) and McCulloch (1994) use the stonecat *Noturus flavus* as a model invader to determine interaction characteristics with the longnose dace *Rhinichthys cataractae*. Wain (1993) and Remnant (1991) both address the potential impact of rainbow smelt on Lake Winnipeg fisheries. The presence of an invasive species may not be detected until it is well established, it may be difficult to remove, and its erradication may not be feasible given current technology and resources.

Economic consequences

A species invasion may directly impact an economic system by causing economic hardship directly or by impacting the outputs of indigenous species. An example of an aquatic invasion that causes direct hardship is the invasion of the Great Lakes by the zebra mussel *Dreissena polymorpha*. The zebra mussel arrived in the Great Lakes aboard an ocean going freighter which expelled its ballast water from Europe into the Great Lakes (Mills et al. 1993). Natural predators and pathogens keep the zebra mussel in check in their indigenous habitat in Europe, but they proliferated in the Great Lakes. Its presence and expansion are creating impacts that require an increase in expenditures by public and private enterprises to repair infrastructure fouled by the mussel. Redesign of power plants has been estimated at $800 million with annual maintenance costs at $60 million. The direct impacts of the zebra mussel to North American waters is expected to reach $3.1 billion (1991 dollars) by the year 2000 (Office of Technology Assessment 1993).

Annual impacts of the ruffe on the Great Lakes fishery, if it enters all of the lakes, has been "conservatively" estimated to reach approximately $7 billion (Great Lakes Fishery Commission 1992). Estimates of the potential impacts were made possible by using data collected locally, and through European experiences with the fish.

Introductions also impact local economies since fewer anglers mean fewer secondary economic impacts in the wider economy. Whereas, a direct impact may affect sport and commercial angling, secondary impacts arise when direct beneficiaries of the resources curtail their use of supporting services (e.g., hotels, resorts, sports stores).

Much of the concern about biota transfer related to the Garrison Diversion Unit (GDU) is the result of incomplete information. It is possible to make predictions based on the best information available and experience of scientists. The best information available at this time is based on only one or more case studies; and for most generalizations, there are a host of exceptions (Ruesink et al. 1995). An example can be found in the basic information available on the natural ranges of many aquatic species, which can often be incomplete or inconsistent. Lack of baseline data sometimes make the distinction between a non-indigenous species and endemic species difficult. For example, a comprehensive survey of the fishes of the Red River system was conducted in order to fill some of the basic gaps in understanding fish ranges. This is summarized in Chapter 6. Heuring (1993) conducted an historical assessment of fish harvests in Lake Winnipeg which provides the first baseline against which to compare future fisheries information. Also, the National Water Quality Assessment Program in the U.S., began in 1991 for the Red River, will provide valuable information on the aquatic biology of the basin (Stoner and Lorenz 1996).

Either positive or negative economic impacts, or both, may result from aquatic introductions. Positive impacts include expanded gamefish resources and biological control of insects, plants, or other fish. Negative impacts may occur when resources, natural environments, or man-made infrastructures deteriorate. This may result in either fishery degradation, efforts to control an introduced species, and ecosystem service deterioration.

Aquatic introductions can have positive economic impacts by providing services and expanding resources in altered habitats such as impoundments and tailwaters. Introduced grass carp *Ctenopharyngodon idellus* serve in many countries to reduce unwanted plant cover in lakes and waterways. Game fish introductions are usually considered positive. Between 25 and 50 percent of ". . . fresh water fishes caught by anglers in the continental United States are from populations established through introductions" (Moyle et al. 1986, p.422). With approximately $24 billion total expenditures by anglers in 1991 in the United States (U.S. Department of the Interior 1993), the motivation for fishery managers to introduce new aquatic species appears high. Most "successful" introductions have occurred in waters which have already been substantially altered by human activity, such as artificial impoundments and tailwaters (Herbold and Moyle 1986: Moyle et al. 1986).

Policy concerns

Central to discussions surrounding species invasion is the role of policy in addressing potential arrival and subsequent impacts of an invasive species. Since there is uncertainty surrounding the potential impacts of many possible aquatic invaders, it is difficult to quantify eventual benefits of any policy alternative and, subsequently, identify a level of effort or spending which should be applied in order to exclude a possible invader. Additionally, it is difficult to isolate the impacts of species invasion from natural variation, from other perturbations (e.g., global climate change), or from unknown sources.

It is difficult to address a potential invasion when there is uncertainty surrounding the mechanism of introduction and the ultimate impacts. Application of policy efforts to a single mechanism of introduction may not be efficient or sufficient. It may be inefficient, or insufficient, to apply policy to a single mechanism of introduction when the potential of introduction through another mechanism continues unabated.

Similarly, the administrative and regulatory frameworks are not entirely in place to address specific invaders. Regulations and restrictions may exist to exclude a potential invader from arriving via an identifiable route, but it may just as easily arrive through an equally viable, yet circuitous route. For example, California restricts the import of fruits and vegetables by mail from tropical countries, but some of the same plant diseases and parasites may still arrive through the mail from Puerto Rico, or be hand carried (OTA, 1993).

One way policymakers can address the problems associated with species invasion is through education. These efforts are usually directed toward aquatic resources users such as anglers and boaters. In a pamphlet which accompanies each angling license, space is devoted to educating anglers about the effects of species introductions and the measures they should take in order to ensure that it does not occur. For example, one page of the 4-page

North Dakota Fishing Regulations, which accompanies each license sold, illustrates the effects of bait bucket release and subsequent effects.

The amount of effort expended on educating the public about aquatic introductions varies from state to state. Gunderson (1995) compared and assessed the effectiveness of the educational strategies of three states, Minnesota, Wisconsin, and Ohio to determine the change in behavior of anglers towards the spread of exotic organisms. The study found that the relatively more aggressive campaign of Minnesota was more effective at changing boater behavior.

The solution to these problems and concerns is simple, yet not politically expedient. Halting the international trade of biological products would certainly reduce the risk of both unwanted target and non-target organisms becoming established in a receiving country. Since this is not likely to be implemented, other solutions need to be explored. There is optimism though, that regulation, education, and controls may altogether reduce the impacts of expansion of potential invaders.

References

Courtenay, W.R., Jr. 1993. Biological pollution through fish introductions. In *Biological Pollution: The Control and Impact of Invasive Exotic Species*, (ed.) McKnight, B.N., 35-61. Indianapolis: Indiana Academy of Science.

Courtenay, W.R., Jr. and J.N. Taylor. 1984. The exotic ichthyofauna of the contiguous United States with preliminary observations on international transplants. Documents presented at the *Symposium on Stock Enhancement in the Management of Freshwater Fisheries. European Inland Fisheries Advisory Commission (EIFAC), Volume 2*, 466-487. Rome: Food and Agricultural Organization of the United Nations.

Crosby, A.W. 1986. *Ecological Imperialism: The Biological Expansion of Europe, 900-1900.* New York: Cambridge University Press.

Elton, C. S. 1958. *The Ecology of Invasions by Animals and Plants.* London: Methuen.

Great Lakes Fishery Commission. 1992. *Ruffe in the Great Lakes: A Threat to North American Fisheries.* Ann Arbor, Mich.: Ruffe Task Force, Great Lakes Fishery Commission.

Gunderson, J. 1995. Three-state exotic species boaters survey: What do boaters know and do they care? *Aquatic Nuisance Species Digest* 1(1):8-10.

Herbold, B. and P.B. Moyle. 1986. Introduced species and vacant niches. *American Naturalist* 128(5):751-760.

Heuring, L. 1993. *A Historical Assessment of the Commerical and Subsistence Fish Harvests of Lake Winnipeg.* M.S. thesis, Winnipeg: The University of Manitoba.

Levin, S.A., 1989. Analysis of risk for invasions and control programs. In *Biological Invasions: a Global Perspective*, eds. Drake, J.A., and H.A. Mooney, 425-435. New York: John Wiley & Sons.

Mills, E.L., J.H. Leach, J.T. Carlton, and C.L. Secor. 1993. Exotic species and the integrity of the Great Lakes. *Journal of Great Lakes Research* 19(1):1-54.

Moyle, P. B., H. W. Li, and B. Barton. 1987. The Frankenstein effect: Impact of introduced fishes on native fishes of North America. In *The Role of Fish Culture in Fisheries Management*, (ed.) Stroud, 415-426, Bethesda, Md.: American Fisheries Society.

Office of Technology Assessment. 1993. *Harmful Non-indigenous Species in the United States.* Washington, D.C.: United States Congress, OTA-F-565, GPO.

Remnant, R.A. 1991. *An Assessment of the Potential Impact of the Rainbow Smelt on the Fishery Resources of Lake Winnipeg.* M.S. thesis, Winnipeg: The University of Manitoba.

Ruesink, J.L., I.M. Parker, M.J. Groom, and P.M. Kareiva. 1995. Reducing the risks of nonindigenous species introductions. *BioScience* 45(7):465-477.

Stoner, J.D. and D.L. Lorenz. 1996. *National Water-Quality Assessment Program: Data Collection in the Red River of the North Basin, Minnesota, North Dakota, and South Dakota, 1992-95.* U.S. Geoglocial Survey Fact Sheet FS-172-95.

United States Department of the Interior. 1993. *1991 National Survey of Fishing, Hunting, and Wildlife Associated Recreation.* Fish and Wildlife Service and the United States Department of Commerce, Bureau of the Census, Washington, D.C.: GPO.

Wain, D.B. 1993. *The Effects of Introduced Rainbow Smelt (Osmerus mordax) on the Indigenous Pelagic Fish Community of an Oligotrophic Lake.* M.S. thesis. Winnipeg: The University of Manitoba.

Welcomme, R.L. 1992. A history of international introductions of inland aquatic species. In *Introductions and Transfers of Aquatic Species,* Sindermann, (eds.) C., B. Steinmetz, and W. Hershberger, *International Council for the Exploration of the Sea* 194:3-14.

CHAPTER 6
Distribution and dispersal of fishes in the Red River basin

Todd M. Koel and John J. Peterka

Executive summary

From surveys made between 1897 and 1994, 77 native and 7 introduced fish species were reported in the streams of the Red River of the North basin (RRNB) in the United States. Of the introduced species, only the common carp has achieved widespread distribution and abundance. White bass, introduced in North Dakota in 1953, gained access to Lake Winnipeg via the Sheyenne River and the Red River. Interchanges of fishes via natural water connections are possible between the RRNB and the Mississippi River basin. Some tributary streams such as the Otter Tail River in Minnesota have several species not found elsewhere in the basin; the large scale stoneroller was collected only in the Forest River in North Dakota. Three native species (longnose gar, lake sturgeon, and pugnose shiner) have not been collected in the RRNB in the United States in recent years, and it is likely their populations no longer exist there.

Introduction

Research objectives

The primary objective of this survey was to provide an accounting of all records of fish species collected from the Red River of the North (Red River) in the United States that have been reported in the published literature. From these records, it was possible to determine fish distribution patterns in the Red River and in 26 major tributaries. Even with a long history of surveys to document fish species occurrence in the Red River basin, there is no detailed treatment of fish distributions in basin streams.

Details of collection sites and dates of collection of fishes in the Red River basin have not been presented in past reports, and will be useful for assessing any changes in fish distributions and for clarifying questions about past records. Further, there is disagreement on the number of fish species in the Red River basin. Crossman and McAllister (1986) listed 75 fish species for the Red River drainage (portions in Canada and the United States); 69 species were listed for the Red River in the United States. Underhill (1989) listed 80 species for the Red River drainage in the United States. The data presented in this report provide the basis to permit various statistical analyses to ascertain environmental variables that are important in influencing fish community assemblages in streams of the Red River drainage.

A second objective of this survey was to determine the current status of the gizzard shad *Dorosoma cepedianum* in the James River and of the white bass *Morone chrysops* in the James River (at Ludden and Lamoure, N.D.) where VanEeckhout collected the gizzard shad in 1989 (Duerre 1989), and in the Sheyenne River in 1993 and 1994 from locations near Valley City to its mouth.

Fishes in the Red River of the North basin

The first published survey of fishes in the Red River and its tributaries in eastern North Dakota and western Minnesota was by Woll (1896). Other early Minnesota surveys by Cox (1897) and Surber (1920) provided several references to fishes in the Red River, and Hankinson (1929) surveyed fishes of the Red River and its tributaries in North Dakota. Many other surveys of fishes of the Red

River basin were consulted in this study (Table 6.1). In addition to information found in published reports, records were obtained from the Minnesota Department of Natural Resources (MNDNR 1994) at Bemidji and Detroit Lakes and from the Ecological Services Section of MNDNR; the North Dakota Game and Fish Department (1994); the North Dakota State Department of Health (1994); National Water Quality Assessment (NAQWA) for the Red River basin; museum records from the University of Michigan (1994); the Bell Museum of Natural History (University of Minnesota) (1994); the Smithsonian Institution (1994), and master's theses from the University of North Dakota, Grand Forks (Copes 1965; Hegrenes 1992; and Russell 1975), and the University of Minnesota. In 1994, sites were seined in the Rush River in North Dakota to provide additional records of fishes and in the Clearwater River drainage in Minnesota to verify the occurrence of the mottled sculpin. No gizzard shad or white bass were found in connection with samples taken as a part of this effort, resulting in the need to rely on sampling conducted by other investigators to document the occurrence of these two species. Detailed results of the RRNB surveys are available in the Ph.D. dissertation of Dr. Todd Koel (1997). Results are also available on the web (Peterka and Koel 1996).

Results

For each species reported in the Red River drainage in the United States (Minnesota, North and South Dakota), documentation is provided regarding dates and location where it was collected; a brief accounting of its relative abundance and habitat requirements; and its current distribution in drainages adjacent to the Red River basin in Minnesota, South Dakota, and North Dakota (Missouri River drainage), and in the Red River drainage in Manitoba. Common scientific names are from Robins et al. (1991). Hybrid fishes, mostly of centrachids, reported in the literature review are not included. The goldfish *Carassius auratus*, reported from the Sheyenne River near Lisbon, and the golden orf *Leuciscus idus*, reported from the Red and Buffalo Rivers, were not included in the list of species in the Red River basin as these were likely introducted from aquaria that have not been able to establish populations in the basin.

Where sufficient information was available, the percent of the total sites sampled in the basin in which each species occurred was used to indicate whether it was rare, or common, and the percent occurrence in sites samples in various ecoregions was used to indicate regions within the basin where it is most successful. These percentages were determined from surveys conducted since 1962 to help include sampling that took place in the entire Red River basin. Ecoregions were defined by Omernik and Gallant (1988). Ecoregions are generally considered regions of relative homogeneity in ecological systems or in relationships between organisms and their environments. Parameters for delineating homogeneity are soils, land use, land surface form, and potential natural vegetation (Omernik and Gallant 1988).

Status of gizzard shad and white bass

There have been no records of the gizzard shad in the James River since a few adults and young-of-the-year were collected near Lamoure in 1988 by VanEeckhout (Duerre 1989 and Power 1995). White bass continue to be reported in the Sheyenne River and the Red River, but they apparently have been unable to establish large populations in the Red River basin since first reported in the Sheyenne River in 1964 from white bass introduced into Lake Ashtabula in 1953. The well established populations of white bass in Lake Ashtabula Reservoir on the Sheyenne River serve as a source for white bass recorded in the Sheyenne River and adjacent reaches of the Red River.

Summary

From surveys made in streams in the Red River basin from 1892-1994, 84 fish species in 20 families were reported; 77 species are now considered native, and 7 are known introductions. The introduced species are rainbow trout, brown trout, brook trout, muskellunge, white bass, common carp, and flathead chub. Of these, only the white bass and common carp have been able to maintain populations through natural reproduction; the only record of the flathead chub was reported from the Red River south of Grand Forks in 1984 (Renard et al. 1985), and it may have entered the Red River

Table 6.1. Sources of fish survey data used to produce species distribution maps.

Streams are the Red (1), Pembina (2), Tongue (3), Park (4), Forest (5), Turtle (6), Goose (7), Elm (8), Rush (9), Maple (10), Sheyenne (11), Wild Rice, North Dakota (12), Bois de Sioux (13), Mustinka (14), Rabbit (15), Otter Tail (16), Pelican (17), Buffalo (18), Wild Rice, Minnesota (19), Sandhill (20), Red Lake (21), Clearwater (22), Snake (23), Middle (24), Tamarac (25), Two (26), and Roseau (27) rivers. Sampling was conducted during Year(s), at various Sites.

	Source[a]	Stream	Year(s)	Sites
1	Bell Museum of Natural History, University of Minnesota	13,14,15,16,17,18,19,20, 21,22,23,24,25,26,27,28	1955-1979	292
2	Copes, F.A. and R.A. Tubb	2,3,4,5,6,7,8,9,10,12	1964	62
3	Enblom, J.W.	27	1976	14
4	Feldman, R.M.	5	1962	8
5	Hankinson, T.L.	1,2,4,11	1922	8
6	Hanson, S.R. et al.	16	1980	14
7	Illinois Natural History Survey	6	1978	1
8	Kreil, R.L. and L.F. Ryckman	2	1987	15
9	MN Department of Natural Resources, Ecological Services Section[b]	1,2,3,4,5,6,7,10,11,13, 14,15,16,17,18,19,20,21, 22,23,24,25,26,27	1983-1994	164
10	MN Department of Natural Resources, Section of Fisheries	16,21,22,24,25,27	1975-1992	53
11	Naplin, R.L. et al.	19	1976	10
12	ND Game and Fish Department	1,8,9,11,12,13	1976-1989	76
13	ND Department of Health[b]	1,2,3,4,5,6,7,8,9,10,12,13	1993-1994	33
14	North Dakota State University, Department of Zoology	9,11,12,18,20,22	1993-1994	33
15	Olson, T.A.	18,19,20,21,22,23	1932	18
16	Peterka, J.J.	11	1977	12
17	Peterka, J.J.	2,3,4,5,6,7	1991	48
18	Renard, P.A. et al.	21	1976-1977	26
19	Renard, P.A. et al.	1	1984	41
20	Russel, G.W.	11,12	1974	38
21	Tubb, R.A. et al.	11	1964	25
22	Wilson, H.W.	11	1950	9
23	Woolman, A.J.	1,2,3,4,5,6,7,10,11,14,16, 18,21	1892	18
24	University of Michigan, Museum of Zoology[c]	6,10,11,16,21	1892-1951	9

[a]See References
[b]Several sites were sampled during a cooperative effort by the Minnesota Department of Natural Resources, Minnesota Pollution Control Agency, North Dakota Department of Health, U.S. Environmental Protection Agency, and U.S. Geological Survey.

from Manitoba. The longnose gar has not been reported from the Red River basin since Woolman's (1896) account, and the last record of the lake sturgeon in the Red River basin in the United States was in the 1950s from Red Lake, Minnesota (Robert Strand, Minnesota Department of Natural Resources, Bemidji, personal communication 1994).

The list provided by Crossman and McAllister (1986) did not include the following species documented in this analysis in the Red River basin in the United States: bowfin, northern hogsucker, largescale stoneroller, common carp, yellow bullhead, central mudminnow, largemouth bass, rainbow darter, rainbow trout, brown trout, logperch, bigmouth buffalo, brook trout, pugnose shiner, green sunfish, and the mottled sculpin. The following species were reported in the Red River basin in the United States by Crossman and McAllister (1986) for which this study was unable to find any documentation that they have occurred in the basin: northern brook lamprey *Ichthyomyzon fossor*, bullhead minnow *Pimephales vigilax*, longear sunfish *Lepomis megalotis*; likewise, we found no documentation that two species listed with a question mark, the silvery minnow *Hybognathus nuchalis* and longnose sucker *Catostomus catostomus*, have occurred [in the basin in the United States.] Underhill's (1989) list of fishes in the Red River basin in the United States did not include the following species that we were able to document: yellow bullhead, muskellunge, orange spotted sunfish, and mottled sculpin.

Compared with other large streams in the region, diversity of fishes in the Red River basin is high, and most of its species are also found in streams of the Mississippi River drainage. The upper Mississippi River (above St. Anthony Falls in Minneapolis) has 69 fish species (Underhill 1989) of which 62 are shared with the Red River. The Minnesota River has 88 species of which 72 are shared. The Missouri River in North Dakota has 65 species (Ryckman 1981) of which 46 species are shared.

Several species are apparently restricted to specific habitats available in only some streams. Species typical of only eastern, clearwater tributaries of the Red River basin are chestnut lamprey, silver lamprey, hornyhead chub, pugnose shiner, blackchin shiner, central mudminnow, and mottled sculpin. Species reported only from the Otter Tail and Pelican River drainages are bowfin, northern hogsucker, central stoneroller, weed shiner, yellow bullhead, rainbow darter, and least darter. The largescale stoneroller has been reported only from the Forest River, and the orange spotted sunfish is most common in the Sheyenne River.

References

Bell Museum of Natural History. 1994. Unpublished fish survey data for the Red River basin. Bell Museum of Natural History, St. Paul: University of Minnesota.

Copes, F.A. 1965. *Fishes of the Red River Tributaries in North Dakota.* Master's thesis. Grand Forks: University of North Dakota.

Copes, F.A. and R.A. Tubb. 1966. *Fishes of the Red River Tributaries in North Dakota.* Contributions of the Institute for Ecological Studies number 1. Grand Forks: University of North Dakota.

Cox, U.O. 1897. *A Preliminary Report on the Fishes of Minnesota.* University of Minnesota, Zoological Series III. St. Paul: The Pioneer Press Co.

Crossman, E.J. and D.E. McAllister. 1986. Zoogeography of freshwater fishes of the Hudson Bay drainage, Ungava Bay and the Arctic Archipelago. In *The Zoogeography of North American Freshwater Fishes*, pp. 53-104, Hocutt, C.H. and E.O. Wiley (eds.). New York: John Wiley and Sons.

Duerre, D.C. 1989. *Ecological Investigations of Lakes, Rivers, and Impoundments in North Dakota (surveys).* Statewide fisheries investigations. Report number A-1157. Bismarck: North Dakota Game and Fish Department.

Hankinson, T.L. 1929. Fishes of North Dakota. *Papers of the Michigan Academy of Science, Arts, and Letters* 10:439-460.

Hegrenes, S.G. 1992. *Age, Growth and Reproduction of Channel Catfish in the Red River of the North.* Master's thesis. Grand Forks: University of North Dakota.

Illinois Natural History Survey. 1994. Unpublished fish survey data for the Red River basin. Champaign: Illinois Natural History Survey.

Koel, T.M. 1997. *Distribution of Fishes in the Red River of the North Basin on Multivariate Environmental Gradients*. Master's thesis. Fargo: North Dakota State University.

Kreil, R.L. and L.F. Ryckman. 1987. A fisheries inventory of the Upper Pembina River in North Dakota. *Prairie Naturalist* 19:121-127.

Minnesota Department of Natural Resources-Ecological Services Section. 1994. Unpublished fish survey data for the Red River basin. St. Paul: Minnesota Department of Natural Resources, Ecological Services Section.

Minnesota Department of Natural Resources. 1994. Unpublished fish survey data for the Red River basin. Bemidji: Minnesota Department of Natural Resources, Section of Fisheries, Region 6.

Naplin, R.L., P.G. Heberling, and H.L. Fierstine. 1977. *Evaluation of Stream Characteristics and Fish Populations of the Wild Rice River Near the Proposed Twin Valley Reservoir, Minnesota*. Special publication number 118. St. Paul: Minnesota Department of Natural Resources, Division of Fish and Wildlife, Ecological Services Section.

North Dakota Game and Fish Department. 1994. Unpublished fish survey data for the Red River basin. Bismarck: North Dakota Game and Fish Department.

North Dakota State University. 1994. Unpublished fish survey data for the Red River basin. Fargo: North Dakota State University, Department of Zoology.

Olson, T. 1932. *Investigation of Fish Life in Certain Streams Tributary to the Red River of the North*. St. Paul: Minnesota Department of Health, Division of Sanitation.

Omernik, J.M. and A.L. Gallant. 1988. *Ecoregions of the Upper Midwest States*. U.S. Environmental Protection Agency EPA/600/3-88/037. Corvallis, Ore.

Peterka, J.J. 1978. Fishes and fisheries of the Sheyenne River, North Dakota. *Proceedings of the North Dakota Academy of Science* 32:29-44.

Peterka, J.J. 1992. *Survey of Fishes in Six Streams in Northeastern North Dakota*. Completion report. North Dakota Game and Fish Department, Bismarck, N.D.

Peterka, J.J., and T.M. Koel. 1996. Distribution and dispersal of fishes in the Red River basin. Report submitted to Interbasin Biota Transfer Studies Program, Water Resources Research Institutue, Fargo, North Dakota. Northern Prairie Wildlife Research Center Home Page. http://www.npwrc.org/resource/distr/others/fishred/fishred.htm.

Power, G.J. 1995. *Ecological Investigations of Lakes, Rivers and Impoundments in North Dakota*. Bismarck: North Dakota Game and Fish Department.

Renard, P.A., S.R. Hanson and J.W. Enblom. 1983. *Biological Survey of the Red Lake River*. Special publication number 134. St. Paul: Minnesota Department of Natural Resources, Division of Fish and Wildlife, Ecological Services Section.

Robins, C.R., R.M. Bailey, C.E. Bond, J.R. Brooker, E.A. Lachner, R.N. Lea, and W.B. Scott. 1991. *Common and Scientific Names of Fishes from the United States and Canada*. Special publication number 20. Bethesda, Md.: American Fisheries Society.

Russel, G.W. 1975. *Distribution of Fishes in North Dakota Drainages Affected by the Garrison Diversion Project*. Master's thesis. Grand Forks: University of North Dakota.

Ryckman, F. 1981. *A Revised Checklist of the Fishes of North Dakota, With a Brief Synopsis of Each Species Distribution Within the State*. Bismarck: North Dakota Game and Fish Department.

Smithsonian Institution. 1994. Unpublished fish survey data for the Red River basin. Washington, D.C.: Smithsonian Institution.

Surber, T. 1920. *A Preliminary Catalogue of the Fishes and Fish-like Vertebrates of Minnesota*. Appendix to the Biennial Report of the State Game and Fish Commissioner of Minnesota for the period ending June 30, 1920. Minneapolis, Minn.: Syndicate Printing Co.

University of Michigan, Museum of Zoology. 1994. Unpublished fish survey data for the Red River basin. Ann Arbor: University of Michigan, Museum of Zoology.

Underhill, J.C. 1989. The distribution of Minnesota fishes and late Pleistocene glaciation. *Journal of the Minnesota Academy of Science* 55:32-37.

Woolman, A.J. 1896. Report upon ichthyological investigations in western Minnesota and eastern North Dakota. Appendix 3. Extracted from the report to the U.S. Commissioner of Fish and Fisheries for 1893. Washington, D.C.: GPO.

CHAPTER 7
Selected case histories of fish species invasions into the Nelson River system in Canada

Kenneth W. Stewart, William G. Franzin, Bruce R. McCulloch, and Gavin F. Hanke

Executive summary

A total of 14 species of fish have been deliberately introduced into the Hudson Bay drainage since European colonization of the area. An additional four species have been introduced either accidentally or illegally, and three apparently have invaded the drainage naturally. Detailed studies of three species which have invaded the drainage within the last 30 years show that their effects have ranged from minimal to significant and may be either deleterious or beneficial (from a human viewpoint). Changes in these effects and perhaps additional effects probably will appear over time. Additonal deliberate introductions, the Garrison Diversion plan, and the live bait and aquarium trades are identified as potential continuing sources of interbasin biota transfer. The only feasible means of control of future biota transfers is a combination of public education and the cooperative development of inspection and control stations at drainage boundaries rather than political boundaries by the jurisdictions involved.

Introduction

The recent invasion of the Nelson River drainage of Hudson Bay by fishes has been occurring, in human terms, for a very long time. Three distinct time periods are represented. The late-glacial period occurred when Lake Agassiz was still in existence. Species colonized Lake Agassiz from the Mississippi/Missouri River area, the Great Lakes area, and the Bering Refugium (Bering Strait/Alaska area) (fig. 7.1). All of these species now are found well beyond the Nelson River drainage, including the Churchill, southern Hudson Bay and James Bay watersheds, the Athabasca and Mackenzie River systems, and/or the Great Lakes/St. Lawrence system and beyond. Fish species of this colonization period are all cool/cold water tolerant.

The second colonization occurred in the post-glacial period following the drainage of Lake Agassiz and has come mainly from the Upper Mississippi River. Two sources were possible: a direct or axial route from the Minnesota River into the Red River mainstem, or a northeastern dispersal route from Mississippi headwaters east of the Red Lakes area of Minnesota into the Rainy and Winnipeg Rivers. All species using the axial route are tolerant of warm, turbid water. All species believed to have used the northeastern route are found in the Winnipeg River system, but not in the Red River mainstem or its tributaries in Manitoba, except isolated locations in the Assiniboine River watershed. All of these species are intolerant of turbid water, but are warm water tolerant.

The third colonizing group in the post-glacial period arrived by means of both the axial and northeast dispersal routes. All of these species are found in the Winnipeg River above Great Falls, as well as in the Red/Assiniboine systems, but do not appear in the Mackenzie River system, indicating their later arrival in the Nelson River drainage. All species in this group are warm water tolerant, with a wide range of tolerances of other conditions.

The most recent additions to the Nelson River fish fauna include species introduced by humans from outside the drainage (Crossman 1991) and range in expansion by species formerly absent

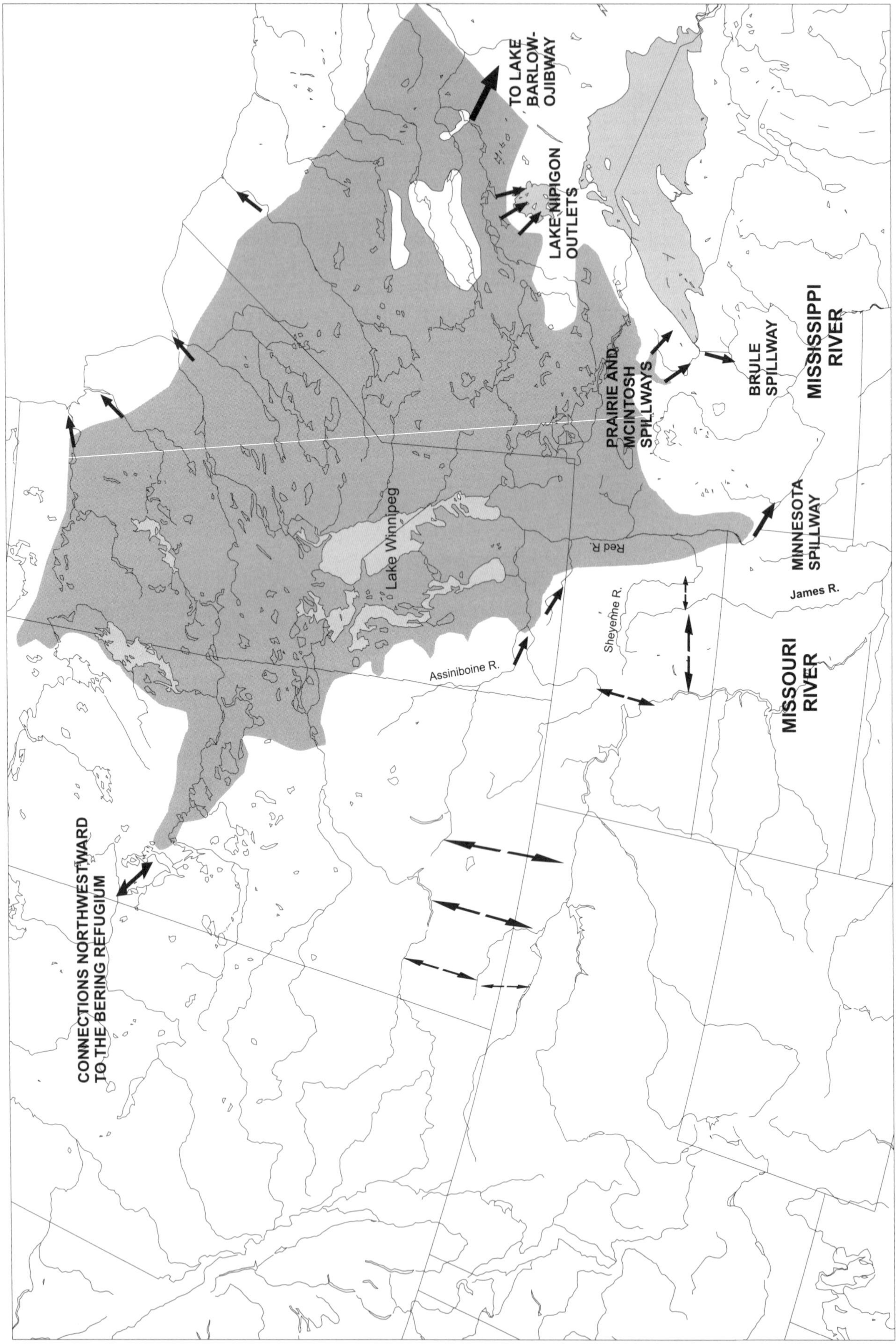

Figure 7.1. Lake Winnipeg within the context of the composite maximum extent of Glacial Lake Aggasiz. At no time did the lake fill the entire shaded area.

from the Manitoba portion of the drainage, but present in headwaters of the Red River drainage in the United States.

Introduction by humans presently is the most frequent way by which exotic fish enter the Nelson River drainage. Deliberate (table 7.1) and unauthorized or accidental release (table 7.2) of game fish, live bait, or other nongame species (Franzin et al.1994; Carlton and Geller 1993) have all contributed new fish species to the Nelson River watershed. Natural dispersal (table 7.3) within the drainage or from headwaters with intermittent or permanent connection to an adjacent drainage system (Stewart and Lindsey 1970; Stewart et al. 1985; Stewart 1988; and McCulloch 1994) have added new species to the drainage and allowed the dispersal of species within the drainage. In addition, illegal release of tropical and temperate aquarium specimens (Nelson and Paetz 1992; Hanke and Stewart 1994) and accidental escape from culture ponds (Atton 1959) also have added exotic species to the drainage.

In recent history, several warm water fish species have entered or have been introduced to the Nelson River tributaries, including the Red and Winnipeg rivers. Of these recent faunal additions, the golden redhorse, bigmouth buffalo, carp, rainbow smelt, and white bass also have successfully entered Lake Winnipeg (Hanke and Stewart 1994). The stonecat has used Lake Winnipeg to disperse to the Brokenhead River (McCulloch 1994) but so far it has not been collected in the lake itself. Similarly, the black crappie has used Lake Winnipeg to disperse up the eastern tributaries, but only rarely is collected in the lake.

There are 96 fish species in the Nelson River drainage in Canada, including 64 in the immediate basin of Lake Winnipeg and its tributaries (57 in Lake Winnipeg), 58 species in the Red and Assiniboine Rivers, and 38 species in Lake Manitoba and its tributaries. There are 44 species in the Saskatchewan River and 69 in the Winnipeg River (Stewart, unpublished data). Eleven species (Koel and Peterka 1994), are known from the Red River and/or its headwaters in North Dakota and Minnesota (Koel and Peterka 1994) but do not occur in the Canadian portion of the drainage (table 7.4).

Concerns about introduced species

The potential effects of introduced stocks or species include alteration of habitat, trophic and spatial relationships of native species, reduced fitness due to reproduction of native stocks with introduced conspecifics, interspecific hybridization, and introduction of disease (Crossman 1991). The stonecat, rainbow smelt and white bass are the only invading species which have been the object of specific studies of impacts in Manitoba. In addition, golden redhorse, spotfin shiner, and goldfish apparently also have spread into Manitoba during the last 10-15 years, but there have been no studies of their impacts on the native fish fauna. The abundance of the spotfin shiner, first detected in Manitoba in 1989, does appear to correlate negatively with abundance of the native river shiner *Notropis blennius* in the Red and Assiniboine Rivers (unpublished data), suggesting a potential interaction between these two minnow species.

Green sunfish *Lepomis cyanellus* occur in the headwaters of the Red River in Minnesota (Koel and Peterka 1994), but have not spread downstream to the Red River mainstem, or into Manitoba up to now.

Red shiners *Cyprinella lutrensis* are common in Lake Francis Case on the Missouri River (Gasaway 1970; Walburg 1977) and a favoured bait species. Interbasin transfer of the red shiner is possible via the live bait trade, as a result of the vague regulations governing use of live bait fish in North Dakota, South Dakota, and Minnesota (Meronek et al. 1995). Red shiners also could be introduced via the tropical aquarium fish trade, which is discussed below. Red shiners show rapid evolution of metabolic compensation for cooler temperatures, even in southern limits of their North American range (King et al. 1985; Zimmerman and Richmond 1981). They are habitat generalists, commonly inhabiting turbid water (Jennings and Saiki 1990; Matthews 1985), and probably will survive if released into southern tributaries of the Hudson Bay drainage. Red shiners are aggressive and have been shown to displace native species in other areas where they have been introduced (Rinne 1991; Douglas et al. 1994). They must be viewed as a threat to the cyprinid fishes of the Red and Assiniboine Rivers.

Table 7.1. Deliberate Introduction of Fish into the Red River Drainage

SPECIES	STATUS	INVASIVENESS
Esox masquinongy muskellunge	1. Native, Winnipeg R., extreme eastern Manitoba 2. Introduced, Duck Mountain Provincial Park, Manitoba	Low. No downstream spread in Winnipeg R., no spread of introduced populations, several failed introductions historically.
Salvelinus fontinalis brook trout	1. Native, Hudson Bay Coastal Plain 2. Widely introduced in Southern Manitoba.	Low. No spread of introductions, little or no natural reproduction in most introduced populations.
S. namaycush X S. fontinalis hybrid "Splake"	Introduced in several lakes in Whiteshell and Duck Mountain Provincial Parks.	Low. No reproduction or spread beyond points of introduction.
Salmo trutta brown trout	Widely introduced in Southern Manitoba	Low. No known reproduction of introduced stocks and no spreading.
Oncorhynchus clarki cutthroat trout	1. Native, Saskatchewan R. headwaters 2. Introduced, Whiteshell Provincial Park	Low. No natural reproduction and no spreading known, previous failed introductions.
Oncorhynchus mykiss rainbow trout	Widely introduced, Southern and Central-Western Manitoba	Low. Little or no natural reproduction in most stocks, no spreading known.
Oncorhynchus nerka kokanee *salmon*	Introduced in Duck Mountain Provincial Park and Lake Winnipeg.	Low. No natural reproduction or spreading known. All known introductions have failed.
Cyprinus carpio common carp	First introduced into Manitoba from 1885 to 1889, apparently unsuccessful (Atton, 1959); first officially recorded in lower Red River in 1938 (Hinks, 1943).	High. Carp have spread to all accessible reaches of the Red and Assiniboine River watersheds, the Manitoba great lakes and down the Nelson River at least to Cross Lake (Atton, 1959).
Morone chrysops white bass	Introduced into Ashtabula Lake, North Dakota (1953) (Koel and Peterka, 1994).	High. White bass have spread to, and are breeding in, much of Lake Winnipeg. They remain low in abundance in the Red and lower Assiniboine rivers.
Morone saxitilis striped bass	Introduced into Devil's Lake, North Dakota	Unknown. Introduced into internal drainage basin with no outlet. Introduction apparently has failed.
Micropterus dolomieui smallmouth bass	Introduced into Winnipeg River and Dauphin Lake watersheds, Lake Athapapuskow and Duck Mountain Provincial Park	Moderate. Has spread downstream in Winnipeg River to Traverse Bay, Lake Winnipeg, and in Dauphin Lake watershed to Dauphin Lake.
M. salmoides largemouth bass	Introduced into Lake of the Woods, (Winnipeg R. watershed), Lake Minnewasta and Canada La Farge clay pit (Red R. watershed)	Low to Moderate. Reproduces where introduced, but no evidence of spread beyond points of introduction except Lake of the Woods where it is now found west to Big Traverse Bay, Manitoba.
Pomoxis nigromaculatus black crappie	Introduced at many places in Winnipeg, Rainy and Red River watersheds, ON, MB, ND and MN.	High. Has spread north to Poplar River on east side of North basin of Lake Winnipeg.
Stizostedion lucioperca zander	Introduced into Spiritwood Lake, North Dakota	Unknown. Introduced into internal drainage basin with no outlet. Introduction may have failed.

Table 7.2. Accidental and/or Unauthorized Introductions: (Escape from Captivity, Bait Fish Release)

SPECIES	STATUS	INVASIVENESS
Amia calva bowfin	1. Native, Ottertail R. headwaters (Koel and Peterka, 1994) 2. Unauthorized release of live bait, Lake of the Woods, 1984 (Crossman 1991)	Apparently low. No downstream spread of Ottertail River fish. Occasional adults taken from Lake of the Woods by the commercial fishery.
Leuciscus idus golden orf	Failed introduction into Red and Buffalo Rivers. (Koel and Peterka, this volume)	Apparently low, introduction has failed.
Osmerus mordax rainbow smelt	Introduced in several lakes in the Rainy, English, and Wabigoon River watersheds in late 1970s and early 1980s (Franzin, et al., 1994).	High. Rainbow smelt have spread throughout Lake Winnipeg and downstream in the Nelson River to the forebay of the Limestone Dam. (Remnant et al. 1997)
Carassius auratus goldfish	Unauthorized introductions; Rock Lake (Pembina R.) Assiniboine R. at Brandon, Red R. at Winnipeg	Moderate. They are reproducing in at least two retention ponds in Winnipeg, and have been reported at several points along the Red River downstream to below the St. Andrews Dam at Lockport, as well as from Rock Lake, on the Pembina River.
Pomoxis annularis white crappie	Probably unintentionally introduced with black crappie	Low. It apparently is reproducing in the Red and/or LaSalle Rivers, where young-of-the-year have been collected rarely from 1982 to the present. It has never been collected elsewhere than the immediate vicinity of the confluence of the LaSalle and Red Rivers.

Table 7.3. Natural Invasion of Hudson Bay Drainage or Expansion of Range Within the Drainage

SPECIES	STATUS	INVASIVENESS
Cyprinella spiloptera spotfin shiner	Established in Red River and tributaries downstream to Lake Winnipeg and in Assiniboine River upstream to Portage la Prairie	High. Had spread to present range by 5 years after first identification in Manitoba in the Roseau River in 1989. Apparently limited at present by Lake Winnipeg and the Assiniboine River Diversion Control Structure at Portage La Prairie.
Moxostoma erythrurum golden redhorse	Established in Red, Assiniboine and lower Brokenhead Rivers, and occurs in the South Basin of Lake Winnipeg.	Moderate. Apparently limited to the west by the Assiniboine Diversion Control Structure. Not abundant anywhere.
Noturus flavus stonecat	Established in the Red and Assiniboine River watersheds from the Brokenhead River west to just below the Shellmouth Dam on the Assiniboine River	Moderate to high. Apparently limited by Lake Winnipeg and by dams on Assiniboine River tributaries built in the early-to mid-1960s, which suggests a mid-to late-1960s entry into Manitoba.

Table 7. 4. Fish Species Found In the Red River and/or its Tributaries In North Dakota and/or Minnesota, but not in the Canadian Portion of the Nelson River Drainage

Species	Common Name	Currently Present?
Lepisosteus osseus	longnose gar	N(?)
Amia calva	bowfin	Y
Campostoma anomalum	central stoneroller	Y
Campostoma oligolepis	largescale stoneroller	Y
Notropis anogenus	pugnose shiner	Y
Pimephales vigilax	bullhead minnow	Y
Leuciscus idus	ide	N
Ictiobus bubalus	smallmouth buffalo	Y
Moxostoma valenciennesi	greater redhorse	Y
Amieurus natalis	yellow bullhead	Y
Lepomis cyanellus	green sunfish	Y
Lepomis humilis	orangespotted sunfish	Y
Etheostoma microperca	least darter	Y

Case histories of invading species

The following three case histories summarize the timing, pattern of spread, and interactions with the resident fish fauna of three invading fish species. The stonecat apparently has spread naturally into the Hudson Bay drainage. The rainbow smelt and white bass are, respectively, unauthorized and intentional introductions.

Stonecat, the case history of a natural invader

Appearance and spread of the stonecat in Manitoba

The stonecat *Noturus flavus* was first collected in Manitoba in 1969 from the Red River, just downstream from the Red River Floodway Control Structure at St. Norbert, Manitoba (Stewart and Lindsey 1970). The present distribution of stonecats in Manitoba (fig. 7.2), and their interactions with native fishes of the Assiniboine River watershed have been studied (Stewart and McCulloch 1990; McCulloch and Stewart 1992; McCulloch 1994).

The dates of closure of dams on the Assiniboine River and on two tributaries of the Assiniboine River, above which stonecats do not occur, support dispersal in the Red/Assiniboine system during the late 1950s and early 1960s. The Shellmouth Dam was completed in 1969 and, although the stonecat has been collected as close to it as 2 km downstream, none have been collected above it. Similarly, the dams on the Souris River at Wawanesa, Manitoba, and on the Little Saskatchewan River at Rivers, Manitoba, were completed in 1952 and 1960, respectively. Stonecats never have been collected upstream of either dam (McCulloch and Stewart 1992).

Stonecats have been collected in Oak Creek, an undammed tributary that joins the Souris River just downstream from the dam at Wawanesa, which demonstrates that the dam probably is a barrier to further upstream movement of stonecats in the Souris River.

Interactions of the stonecat with native fishes

Stonecats were found to inhabit riffles in the Assiniboine River and its tributaries. They were found with a number of fish species, but the longnose dace *Rhinichthys cataractae* was the species most often found with them (McCulloch 1994). McCulloch (1994) conducted several 24-hour collection series at stations above and below the Rivers Dam, which is impassable to stonecats. These studies revealed that stonecats were consistently nocturnal in their feeding activity and that they fed on a variety of aquatic invertebrates. Longnose dace, in contrast, did not show significant diet variation in feeding activity and showed a strong dominance of hydropsychid caddis fly larvae in their stomach contents. Longnose dace were about equally abundant in the downstream and upstream sites.

Juvenile burbot *Lota lota* also occurred with stonecats in the downstream site on the Little Saskatchewan River, and in the absence of stonecats in the upstream site. They were much more abun-

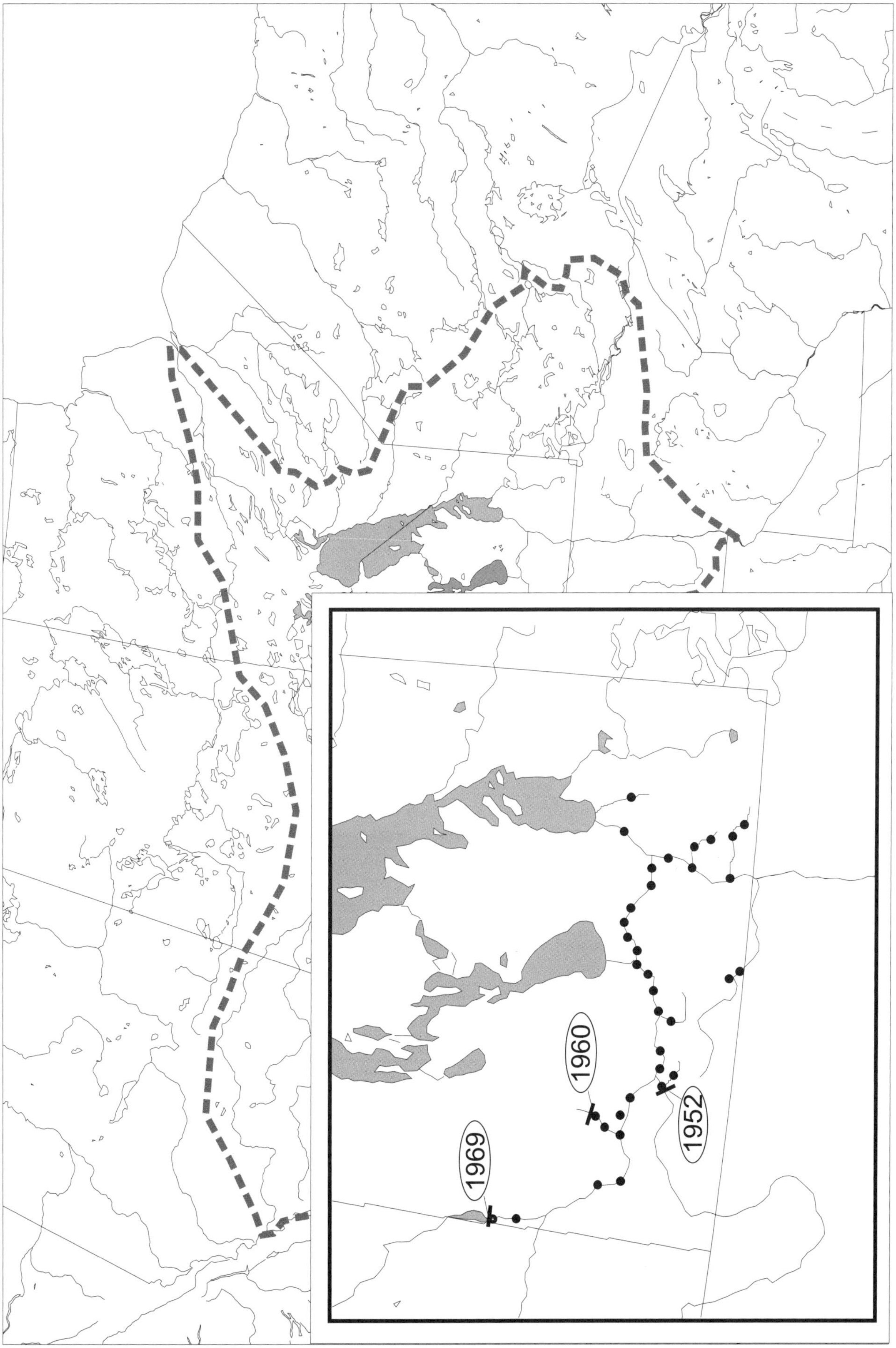

Figure 7.2. Current known distribution of Stonecat (inset) in the Nelson River drainage of Manitoba and Northwestern Ontario. Bars indicate presence of dam with year of construction.

dant at the upstream site. This may be a result of the presence of more pool habitat at the upstream site, since burbot prefer lower water velocities than do stonecats (McCulloch 1994).

Stonecats appear to have reached the maximum extent to which they can disperse in Manitoba, barring human introductions. McCulloch and Stewart (1992) noted that there was little evidence that they had limited the distribution or abundance of other native Manitoba fish species. They further noted that the native species with which the stonecat occurred in Manitoba also occurred with the stonecat in the Upper Mississippi River. All of these species probably entered the Hudson Bay drainage from the Upper Mississippi at various times following deglaciation of the Hudson Bay drainage (Stewart and McCulloch 1990). The stonecat, whether from natural dispersal or human introduction into the Red River, is one of the most recent colonizers among these species. It is likely, then, that the stonecat and the other native species were adapted to co-existence in the same habitats before any of them managed to disperse into the Hudson Bay drainage. If this is true, then competitive interactions between the stonecat and the other species with which it occurs in Manitoba should be minimal, and this appears to be the case.

Rainbow smelt, an unauthorized introduction to the Hudson Bay drainage

Invasions of rainbow smelt

The rainbow smelt *Osmerus mordax* (Mitchill) originally was restricted to anadromous and isolated inland populations in North America mainly east of Appalachia. It has been introduced as a forage species in many lakes west of its original range, beginning around the turn of the century. It is an invasive species which is considered a desirable forage species by some fishery managers and a deleterious exotic by others.

The distribution of rainbow smelt in Ontario resulting from its deliberate introduction from landlocked native populations in Maine into the Great Lakes as early as 1912 has been described, as known in 1986, by Evans and Loftus (1987).

A number of intentional and accidental introductions of rainbow smelt have occurred outside of the Great Lakes basin, particularly in the upper Mississippi River system in a number of states, and in the headwaters of the Rainy River of the Nelson River drainage (Winnipeg River system) in Minnesota and northwestern Ontario.

The following account of the spread of rainbow smelt in the Missouri-Mississippi drainage after a point introduction illustrates the colonizing potential of the species. Rainbow smelt were introduced into a Missouri River reservoir, Lake Sakakawea, North Dakota, from Lake Superior in the spring of 1971 by the North Dakota Game and Fish Department to serve as a forage base for sport fishes (Dyke 1989). A self-sustaining population of rainbow smelt established itself within a few years. Rainbow smelt appeared in Lake Oahe, South Dakota, the first reservoir downstream of Lake Sakakawea by 1974; and large numbers of smelt are now present in both of these Missouri River mainstem reservoirs (Mayden et al. 1987). Rainbow smelt also are found in the next three downstream reservoirs but are not as well established, apparently because of inferior habitat (Mayden et al. 1987). This species also has been recorded from a number of locations further downstream in the mainstem portions of the Missouri and Mississippi rivers and was collected from as far south as Louisiana in 1979 (Suttkus and Connor 1979). Rainbow smelt were collected in the Missouri River upstream of Lake Sakakawea as far as the tailrace of the Fort Peck Dam and in the Yellowstone River, Montana in 1979 (Gould 1981). Most Mississippi and all Missouri River occurrences are presumed to have resulted from the Lake Sakakawea introduction (Mayden et al. 1987). Rainbow smelt found in the Illinois River (and some in the Mississippi River) are assumed to have arrived there from Lake Michigan via the Chicago sanitary canal, but a single specimen taken from the Ohio River in 1986 is of unknown origin (Mayden et al 1987).

The first record of rainbow smelt from the Hudson Bay drainage basin was from Little Eagle Lake, near Dryden, Ontario in about 1962 (Campbell et al. 1991). This small closed-basin lake was treated with poison in 1978, and the population of rainbow smelt was eradicated (Campbell et al. 1991). It is not known whether other populations of rainbow smelt in the basin were established via dispersal from Little Eagle Lake prior to the poison treatment. Rainbow smelt also have been present in the upper portion of the Rainy River system since

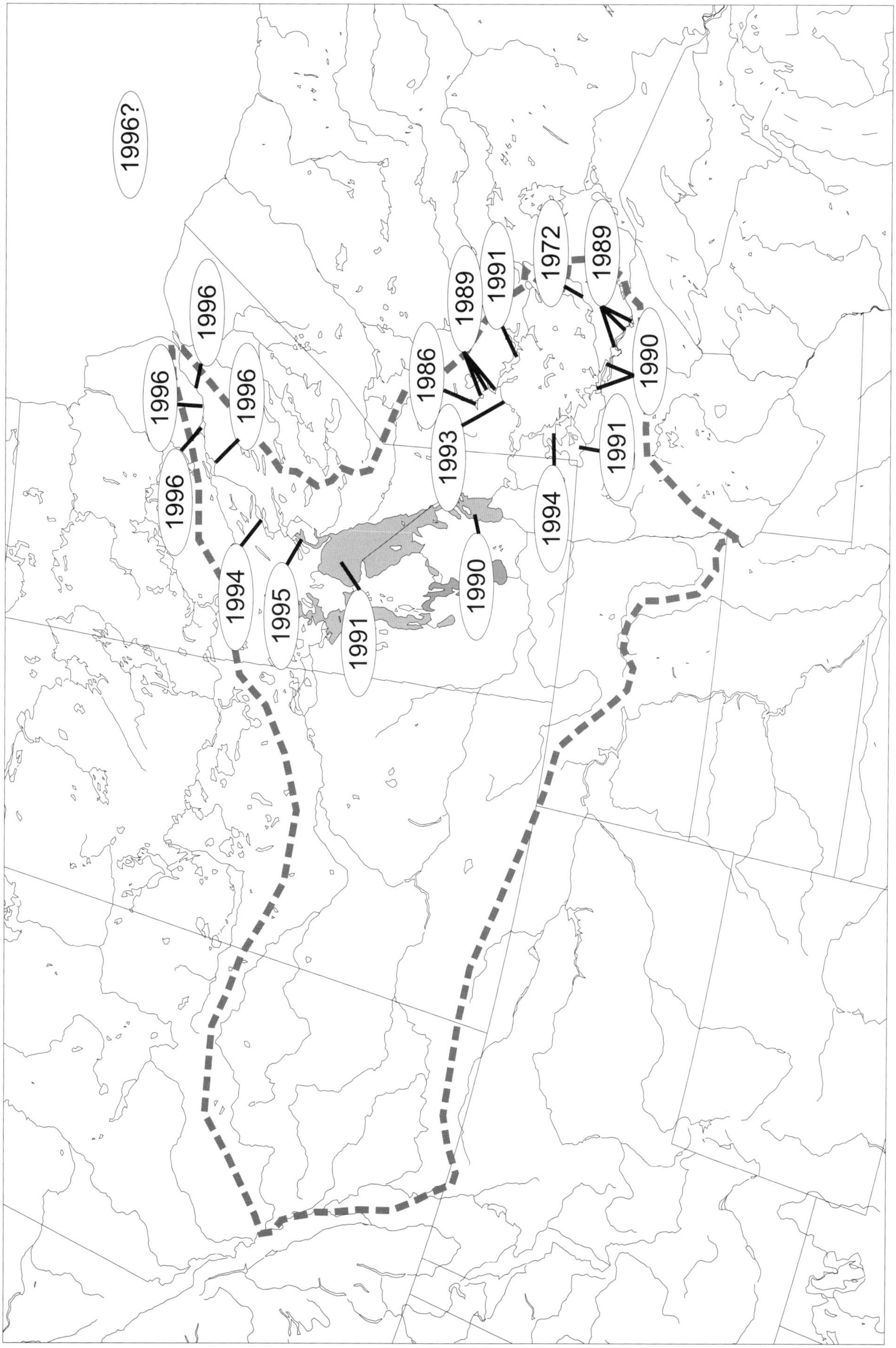

Figure 7.3. Current known distribution of Rainbow Smelt in the Nelson River drainage of Manitoba, Northwestern Ontario, and northeastern Minnesota. Dotted line outlines approximate boundary of the drainage basin.

1972, in Burntside Lake, Minnesota, and Eva Lake, Ontario (Campbell et al. 1991).

Franzin et al. (1994) conducted extensive field and literature surveys in 1989 and 1990 to document the distribution of rainbow smelt in Hudson Bay drainage waters of northwestern Ontario, southeastern Manitoba, and northern Minnesota as it was known in 1991. They examined available ecological information on rainbow smelt, predicted the potential for further spread of this species in the basin, and discussed some possible effects of its colonization. In 1991, rainbow smelt were known from 32 lakes of the Winnipeg River drainage in Manitoba, northwestern Ontario, and northeastern Minnesota, including sixteen new records resulting from the 1989 and 1990 surveys and subsequent reports.

Rainbow smelt have consolidated their positions in the fish communities of many lakes and have spread to several new locations since the 1994 report—in some cases, hundreds of kilometres down major river systems (fig. 7.3). The species has become very numerous in Rainy Lake as indicated by hydroacoustic surveys and midwater trawling (L. Kallemyn, Voyageur National Park, pers. comm.). Rainbow smelt passed down the Rainy River and have become abundant in the main basin of Lake of the Woods, and they are known from most areas of the lake except for Whitefish Bay and Shoal Lake. Spawning runs have been observed on Blindfold and Longbow creeks in the northeast area of the lake (N. Ward, Ontario Ministry of Natural Resources pers. comm.). This evidence suggests that Lake of the Woods now harbours a large, widespread rainbow smelt population. They have not been detected in the Winnipeg River, but were taken from fish screens on turbine intakes at the hydro dam on the outlet of Lake of the Woods at Keewatin in 1995 (T. Mosindy, Ontario Ministry of Natural Resources, pers. comm.).

Lac Seul, a large reservoir on the English River, also has developed a large rainbow smelt population, particularly in the west end of the basin near Ear Falls Dam. Also, Minnitaki, Abram, and Pelican Lakes, which lie between Sandybeach Lake (one of the point introductions reported by Franzin et al. 1994) and Lac Seul, now contain abundant rainbow smelt populations (B. Allen, Ontario Ministry of Natural Resources, pers. comm.). These lakes combined provide a large source of rainbow smelt colonists for lakes downstream in the English River mainstem. The Chukuni River lake chain beginning with Red Lake and ending with Pakwash Lake, which feed into the English River below the Ear Falls Dam, also contains large populations. These two sources could account for the presence of smelt in Maynard Lake and probably also Oak Lake (N. Ward, Ontario Ministry of Natural Resources, pers. comm.). Oak Lake is one of the first lakes in the river below Ear Falls with suitable off-stream habitats for rainbow smelt. The English River joins the Winnipeg River below Lake of the Woods. Since there are two large sources of smelt upstream, populations probably will develop in forebays of hydro dams in the lower English River and along the Winnipeg River in Manitoba.

Rainbow smelt were captured for the first time from Lake Winnipeg in 1990 (Campbell et al. 1991) and now are widespread in the lake. By 1996, the species had spread down the Nelson River to Playgreen, Sipiwesk, and Split Lakes. It had also spread further downstream through Stephens Lake, the forebay of the Kettle Rapids Dam, to the forebays of Limestone and Long Spruce hydro dams on the lower Nelson River about 175 km above Hudson Bay (Remnant et al. 1997). This dispersal is comparable to that seen in the Missouri-Mississippi system. Rainbow smelt can be expected to disperse to Hudson Bay and develop anadromous populations in Hudson Bay and James Bay rivers. It is not known how rainbow smelt arrived in Lake Winnipeg, but live bait release in the Red River or Lake Winnipeg itself may have been involved; the species appeared in Lake Winnipeg without having been observed in the connecting waters between Lake Winnipeg and English and Rainy River watersheds.

Effects of rainbow smelt

Many retrospective studies have been written on the effects of rainbow smelt invasions, most recently those of Evans and Loftus (1987), Evans and Waring (1987), and Nellbring (1989). These studies identified the potential threat posed by rainbow smelt to existing sport and commercial fisheries and attempted to define some of the interactions that have occurred in lakes where rainbow smelt have become established. Evans and

Waring (1987) identified a basic problem; the potential for success of smelt colonization or the possible ecological effects of the species in a lake ecosystem are not always predictable from physical and biological data.

Rainbow smelt are known to have deleterious effects on other fish species, especially in smaller lakes (Evans and Loftus 1987; Evans and Waring 1987; Loftus and Hulsman 1986; Wain 1993), and have been implicated in a reduction in the total value of Lake Erie fisheries (Remnant 1991). Rainbow smelt impacts on important commercial and recreational fisheries already are appearing in recently invaded lakes in northwestern Ontario. Lake whitefish and cisco abundances in Red Lake have declined significantly following the development of a large rainbow smelt population. At the same time, walleye and lake trout growth and abundance have increased markedly in the lake. However, lake trout have developed an "off flavor," and both species are much fatter than they were before they switched to diets of mainly rainbow smelt (R. Wepruck, Ontario Ministry of Natural Resources, pers. comm.). Increased growth of walleye has been reported in other locations, both in northwestern Ontario (Van den Broeck 1995) and in Colorado (Jones et al. 1994), and may well represent an expected outcome of introducing rainbow smelt into walleye lakes. Notwithstanding the potential increased walleye growth, the deleterious consequences of their introduction, both in the stocked lake (negative effects on other species) and in the broader drainage basin should they emigrate, are strong arguments against any further introductions of rainbow smelt.

Even with regulations restricting the use of rainbow smelt as bait, it is reasonable to expect their continued spread to accessible lake trout lakes throughout the populated parts of the Nelson River drainage basin. The invasion of rainbow smelt in the basin presents the opportunity to carry out before and after research to determine definitively the effects of a colonizing fish species in a variety of lake settings. For example, work is in progress to learn effects of rainbow smelt on mercury uptake by lake trout and walleye in lakes in northwestern Ontario. Mercury in tissues of lake trout in Lake of the Woods was determined prior to invasion by rainbow smelt and will be followed over the next few years. Lake of the Woods lake trout had low mercury levels in their tissues prior to rainbow smelt invasion, and levels may be expected to climb as trout turn to a diet comprised mainly of rainbow smelt. Similarly, mercury levels of walleye are known for Rainy and Maynard Lakes and Lac Seul, on the Rainy and English Rivers, respectively, areas with a history of elevated mercury in fish tissues due to reservoir construction. The switch by walleye to diets comprised mainly of rainbow smelt in lakes where mercury levels in walleye tissues already are near the level at which human consumption warnings would be issued may be sufficient to raise those levels above acceptable concentrations.

White bass, an intentional introduction

Introduction and spread of white bass

North Dakota introduced white bass *Morone chrysops* to Lake Ashtabula, a reservoir on the Sheyenne River within the Hudson Bay drainage basin, in 1953 (Koel and Peterka 1994). By 1963, it had moved downstream via the Red River into Lake Winnipeg where it is present now in substantial numbers. During a survey of the eastern tributaries of the North Basin of Lake Winnipeg conducted in 1992 and 1993, young-of-the-year white bass were found as far north as the mouth of the Mukutawa River (53^0 10' N: 97^0 26' W) (Hanke 1996). This is within 67 km of Warren's Landing, at the outlet of Lake Winnipeg. Additional surveying done by Hanke in 1993 and 1994 (Hanke 1996) found white bass northward to the Beaver Creek Provincial Campground (51^0 23' N: 96^0 55' W) on the west side of the Narrows area of Lake Winnipeg (fig. 7.4). No white bass were collected along the western shoreline of the north basin of Lake Winnipeg and there was no evidence of white bass spawning on the western shore of the lake north of the Icelandic River (51^0 00' N, 96^0 37' W).

White bass may not have completed the full colonization of Lake Winnipeg since they are still rare to absent on the west side of the North Basin. They had been present for years in small numbers in the Red River and South Basin of Lake Winnipeg in the early 1980s (Stewart et al. 1985). In the mid-1980s, rapid expansion in numbers and distribution in Lake Winnipeg began, and by the early 1990s, they had attained their present distribution (Hanke

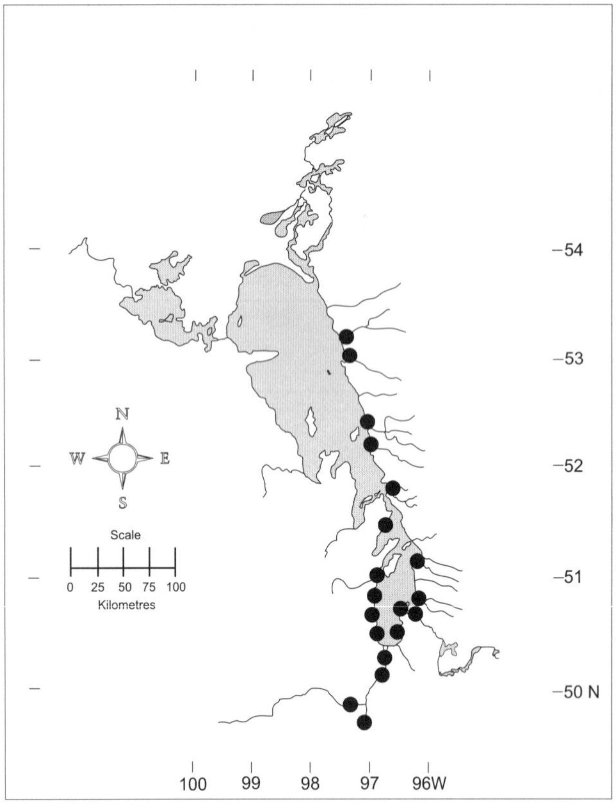

Figure 7.4. Distribution of White Bass (*Morone chrysops*) in Lake Winnipeg and vicinity.

1996). The recent increase in numbers is probably due in part to the warm summer weather during the late 1980s, which would have favored earlier reproduction, increased growth of young-of-the-years and, consequently, more successful overwintering of young-of-the-years.

Effects of white bass

The impact of the introduction of white bass on other species in Lake Winnipeg is unclear because they are predators of the open water pelagic zone, a niche that was previously unoccupied. Young white bass are known to form mixed schools with yellow perch *Perca flavescens* (Hanke 1996), but usually are found in single species schools along the sandy beaches of Lake Winnipeg. There is no evidence for segregation of nearshore habitat between native species and young white bass in the south basin of Lake Winnipeg. Similarly, there appears to be no dietary segregation by nearshore fishes in Lake Winnipeg. The young-of-the-year of several fish species using the nearshore habitat appear to take the same prey species, most commonly calanoid copepods and the cladoceran *Leptodora* sp.; later, as the fish grow, predatory fish switch to either aquatic insects or fish, mainly yellow perch and emerald shiners (Hanke 1996). Emerald shiners are the most widespread and abundant fish in Lake Winnipeg, followed by young-of-the-year yellow perch, and both are common items in the stomachs of Lake Winnipeg piscivores (Hanke 1996). The potential for competition for prey between white bass of all life history stages and representatives of the native fish fauna is minimized by the fact that white bass and most of the native fish feed on the most abundant prey items in the lake. Adult white bass also probably are segregated from other piscivores in Lake Winnipeg by water depth, with white bass occupying the mid to upper water column and walleye and sauger remaining near the bottom of the lake. Because of the abundance of white bass and the fact that all piscivores feed primarily on yellow perch and emerald shiners, the effect of the success of the white bass introduction ultimately may be to cause a decline in the population of these two species. The deep-bodied shape of white bass favors their survival in Lake Winnipeg, since both walleye and sauger, the most abundant native predators in the lake, are gape-limited. White bass 2 years and older, therefore, are nearly free from predation by these species.

Coincident with the arrival of the white bass in Lake Winnipeg, there has been a decline in the landed weight of walleye and sauger from the commercial fishery. Several other factors, such as increased eutrophication of the lake, overfishing, and climatic changes, preclude establishing a direct correlation between the arrival of white bass and the decline in the walleye/sauger fishery.

A possible interaction among emerald shiners, rainbow smelt, and white bass may develop as both smelt and white bass become more abundant. Rainbow smelt are known to be zooplanktivores when young and may switch to some degree of piscivory as adults (Henderson and Nepszy 1989). The addition of rainbow smelt and white bass effectively adds two species which are zooplanktivores when young and piscivores (or potentially piscivorous in the case of smelt) as adults. Therefore, adult smelt and bass will feed on the forage fish species that they competed with when young, such that native forage species are directly or indirectly influenced by these new colonizers throughout their lives. White bass appear to be most successful in dominating the nearshore habitat during warm years, whereas smelt populations can be expected to increase during cooler years. The difference in thermal preference of smelt and white bass ensures that there should always be a substantial population of one or the other of these non-native species in Lake Winnipeg. Because most pelagic fish species rely heavily on three prey taxa (calanoid copepods *Leptodora* spp., emerald shiners, and yellow perch), any disturbance that will adversely affect these important prey items will probably have a cascading effect throughout the entire lake ecosystem.

Varied consequences of introductions

In summary, the three case histories studied to date show that the effects of exotic fish introductions are varied and complicated. The most desirable result following the introduction of an exotic species is no competitive interaction between the exotic and native fish. This result appears to be the case with the stonecat, since it inhabits fast water not frequented by native species except for the longnose dace. It feeds at a different time of the day than longnose dace, however, and on different prey items. The introduction of white bass to the Red River system added a new top predator and a possible competitor with native fish communities of the Hudson Bay drainage. In Lake Winnipeg, the white bass and its potential competitors appear to feed on the same four abundant prey species. The high abundance of these species at present appears to mask potential negative effects of competition following the addition of the new predator. The introduction of rainbow smelt adds trophic complexity to the food chain, and at the same time, a potential competitor with the prey species. The alteration of the food chain may further concentrate pollutants in the tissues of economically important piscivorous fish such as walleye and sauger if they turn to a diet dominated by rainbow smelt with possible damaging effects on an established commercial fishery. Because of competition with and/or predation on native species, either by rainbow smelt, or in addition to white bass, there may be declines of other native species.

Other deliberately introduced species

During the 1980s, North Dakota attempted to introduce a European percid, the zander *Stizostedion lucioperca,* into Spiritwood Lake, a closed basin within the Red River drainage. Canada and Manitoba and concerned North Dakotans mounted very strong opposition to this project since there were very serious potential ecological consequences for Canadian waters and throughout the Missouri-Mississippi basin if zander escaped into waters connected to either the Missouri or Red River drainages. As a result of these efforts, the Governor of North Dakota ordered a halt to the project. The North Dakota Department of Game and Fish subsequently initiated an inter-agency committee, with representatives of the fisheries agencies of Manitoba, Minnesota, Montana, North Dakota, South Dakota, and Saskatchewan, to examine any future fish introductions in the region. The committee agreed that before any new species may be introduced by any jurisdiction, a risk analysis must be undertaken and that the proposed introduction must have the consent of all potentially affected neighboring jurisdictions (Wright and Franzin 1998).

Other sources of exotic biota

Escape from culture

Grass carp *Ctenopharyngodon idella* and bighead carp *Hypophthalmichthys nobilis* have escaped from culture in the lower Missouri and Mississippi rivers. These large, herbivorous cyprinids are both sufficiently tolerant of cool temperatures to make survival in the Nelson River watershed likely. Grass carp eat macrophytes, and may pose a threat to Nelson River native species whose young life history stages use macrophyte beds for shelter. Bighead carp feed on planktonic algae and have the potential of causing complex, bottom-up effects on the entire aquatic food chain of the region.

Grass carp are being cultured in Alberta at present, and a risk assessment for the use of sterile triploid individuals of this species to control aquatic macrophytes in irrigation supply channels and reservoirs within the South Saskatchewan River watershed has been prepared. The possibility of escape of grass carp into the South Saskatchewan River and the less than complete reliability of the process of producing sterile triploid fish pose a threat that grass carp may eventually establish populations in the South Saskatchewan River and subsequently spread downstream. Even if no reproduction occurs, any escaped sterile grass carp may cause substantial damage to existing macrophytes in Saskatchewan River tributaries.

Release of live bait

Among the many issues that have arisen during the development of the Garrison Diversion project, the one that keeps emerging as an uncontrolled source of biota transfer is release of live bait. Large-scale water diversions can create enhanced recreational fishing opportunities. The original 1965 Garrison Diversion plan and a recent plan put forth by the State of North Dakota proposed the creation of a large reservoir known as the Lonetree or Mid-Dakota Reservoir, to be constructed in the headwater area of the Sheyenne River and filled with water diverted from the Missouri River (fig. 7.1). The Missouri and Hudson Bay drainages would have been separated by the width of the reservoir embankment in that area. The creation of a reservoir inevitably leads to the development of a recreational fishery and the potential for transfer of non-native biota by release of live bait. Despite all that engineering can do to eliminate or reduce the risk of interbasin biota transfer with diverted water, the actions of sometimes well-intentioned, but ill-advised, anglers are a major vector for the introduction of non-native species. In addition to the effects of bait fish introductions, there is also the potential for transfer of other non-native species, such as crayfish and leeches used as bait, and zebra mussels, spiny water fleas, Eurasian water milfoil, and pathogens of aquatic biota incidentally transferred with bait species. The effects of these incidental introductions may be as significant as the non-native fish themselves.

Two recent papers examined live bait use: one in Ontario (Litvak and Mandrak 1993) and one in North Dakota (Ludwig et al. 1994). Litvak and Mandrak discovered 28 species of fish, six of which were illegal as bait, in the tanks of four Toronto bait dealers. The fishing destinations of 34 customers interviewed revealed that at least one species of 18 sold was being taken outside of its natural range. More disturbing was the finding that 14 of the 34 anglers were in the habit of releasing live bait fish into the waters they fished in spite of regulations to the contrary. Ludwig et al. (1994) surveyed bait dealers and anglers in the North Dakota/Minnesota area near the boundaries of the Hudson Bay and Mississippi drainages. They found that bait fish wholesalers and retailers were buying and selling bait fish across drainage basin boundaries. Also, anglers were moving across boundaries with both captured and purchased bait fish. Neither Minnesota nor North Dakota has regulations concerning interbasin fish movements but there are regulations on the movement of bait fish among states. Some bait fish caught in the Mississippi basin are transferred to the Hudson Bay basin especially in fishing areas near the basin boundaries. In addition, the advent of boats with livewells capable of holding fish for several days and the increase in popularity of fishing tournaments have increased the probability that bait fish and other biota will be introduced to new ecosystems where they may survive, reproduce, and possibly have deleterious effects. Recent increases in the number of inland Wisconsin lakes adjacent to Lake Michigan that have been invaded by zebra mussel (Wiland 1995) illustrate the potential of this problem.

To date, we have identified no exotic fish species in bait tanks of Manitoba and northwestern Ontario (Kenora and Lake of the Woods area) bait dealers. Dealers' bait tanks usually contain two or more of the following species: finescale dace *Phoxinus neogaeus*, northern redbelly dace *P. eos*, pearl dace *Margariscus margarita,* and fathead minnows *Pimephales promelas*. Less commonly, they have golden shiners *Notemigonus crysoleucas* and blacknose dace *Rhinichthys atratulus*. Brook stickleback *Culaea inconstans* are common contaminants. The golden shiner and the blacknose, pearl, redbelly, and finescale daces have restricted distributions in Manitoba, either in headwater streams or in scattered pockets of suitable habitat. The fathead minnow probably is the most suitable species to offer for sale as live bait because it is widespread and common in most waters in Manitoba.

The only real solution to the problem of biota transport as a result of the use of live bait is a complete ban on the transport of live bait. Such a ban is highly unlikely, given the economic losses that would result from the closure of the bait collection and distribution industry (perhaps $30 million per year in the province of Ontario alone). An intensive educational program for boaters and anglers, and some control over the bait species that are offered for sale, would help to mitigate the problem. The Manitoba Department of Natural Resources recently (1994) has produced a 24-page field guide to the common bait species used in Manitoba in an effort to educate the public. Efforts such as this field guide hopefully will generate concern and public interest in more than just game species and thereby help prevent future degradation of natural fish populations by live bait transfer.

If the transport of live bait is to continue, it is critical for the jurisdictions involved to recognize that inspection and control of transport of aquatic biota must take place at drainage, not political, boundaries. If the jurisdictions could develop a cooperative program of regular inspection of boat livewells, cooler chests, and other equipment being transported by anglers at boundaries between major drainage basins, it may result in a feasible solution to the problem of live bait transfer of aquatic biota.

Release of aquarium fish

The aquarium trade is a recognized source of introduced exotic species in warmer climates such as Florida and the Gulf Coast states. The tropical and temperate aquarium fish species introduced and/or established in North America include, 17 species of cichlids; 8 poeciliids; 5 cyprinids; 3 characids, including the piranha; 2 loricariids; 1 clariid; 1 cobitid; 1 oryziatid; and 3 belontiids (Lee et al. 1980; Crossman 1991). The number of aquarium species already encountered in North America illustrates the potential risk that the aquarium fish trade poses to native North American fish stocks. Of the aquarium fish listed above, 5 cichlids, 4 poeciliids, 3 cyprinids, 2 characids, 1 loricariid and 2 belontiids have been introduced to Canadian waters (Crossman 1991; Nelson and Paetz 1992).

In Manitoba, the goldfish *Carassius auratus* is the only aquarium fish known to be established. A breeding population in two connected storm water retention ponds in southern Winnipeg has existed for at least the last 8 years. There are also sporadic reports of adult goldfish in the Red River downstream as far as the mouth of the Red River floodway at Lockport, Manitoba, and two juveniles were collected from Rock Lake (Pembina River Watershed) in 1993 (specimens collected by B. Yake, Manitoba Dept. of Natural Resources, Fisheries Branch). The retention ponds are intermittently connected to the Red River during periods when runoff from heavy rains or spring melt water drains into the river. Goldfish may use these connections as an avenue for dispersal into the Red River. The other fish currently inhabiting the pond system: carp, fathead minnows, brook stickleback, and the black bullhead must have entered the ponds by upstream movement from the Red River.

Red shiners *Cyprinella lutrensis,* mentioned previously as a high-risk species, are occasionally offered for sale in Winnipeg aquarium shops mixed with shipments of low-priced goldfish. Red shiners also are misidentified in the aquarium trade as 'Asian rainbow barbs' or 'rainbow mountain dace' in shipments of tropical fish and clear customs inspection without suspicion.

The ide *Leuciscus idus,* a common garden pond fish native to Europe, has been recorded from the headwaters of the Red River (Koel and Peterka 1994). The ide also may have been released with live bait, but release from the aquarium fish trade also is likely. Similarly, the tench *Tinca tinca* is known to survive in ponds in British Columbia (Scott and Crossman 1973) and these fish could have been unwanted aquarium fish or released for angling purposes. The tench is still a desired pond species (H. van Elst, pers. comm.) and again, as in British Columbia, may establish breeding populations if released elsewhere in Canada.

In addition to these species, the green sunfish *Lepomis cyanellus* also is sold in the aquarium fish trade, misidentified as a South or Central American cichlid. Channel catfish *Ictalurus punctatus* are sold under the false name 'blue channel catfish', and shovelnose sturgeon *Scaphirhynchus platorhynchus* and Iowa darters *Etheostoma exile* are occasionally available to aquarium enthusiasts. The channel catfish and Iowa darter are native to Manitoba, but there is a risk that introductions from the pet trade may contaminate the local gene pool with input from southern populations or hatchery stock, thereby reducing their fitness. Longnose gar *Lepisosteus osseus* also are commonly imported into Canada, misidentified as 'Florida gar'. The presence of longnose gar in the Great Lakes (Scott and Crossman 1973) and the record from the headwaters of the Red River (Koel and Peterka 1994) suggest that this species also can survive elsewhere in Canada and should not be imported. The bitterling *Rhodeus sericeus* is a European species that is imported and has been identified in Winnipeg pet shops. If introduced, it may survive in the Hudson Bay drainage as it does in sites in the United States (Lee et al. 1980). Popular pond and aquarium fish that are native or established in Europe such as goldfish, carp, and tench all have established populations in southern Canada. In addition, the rudd *Scardinius erthrophthalmus* and weather loach *Misgurnus anguilicaudatus* have established populations in the northern United States (Lee et al. 1980) and must be viewed as high-risk species for accidental and/or intentional introduction to Canadian waters.

Tadpoles of the bullfrog *Rana catesbiana* are a common contaminant transported in bulk shipments of goldfish. These frogs commonly are raised beyond metamorphosis as pets by aquarists and, like goldfish, may be released into natural ponds by well meaning but uninformed people. One of us (Stewart) has received anecdotal reports from cottage owners in the area that attempted introductions of bullfrogs into the Nettley-Libau Marshes at the south end of Lake Winnipeg were made sporadically up to the 1950s. All of these attempts apparently have failed. Bullfrogs have, however, been established in Burnaby Lake, British Columbia, for many years.

The only feasible solution to the risk posed by the aquarium fish trade is to educate the public to the dangers of releasing exotic animals into new environments. Again, a mechanism for inspection of shipments of live aquatic organisms at watershed boundaries, not political borders, is critical to preventing additional introductions.

Conclusions

- Fish colonizing the Hudson Bay drainage from the Mississippi and perhaps Missouri headwaters have evolved in the presence of those species already present in the Hudson Bay drainage. They may re-create the same niche as they occupied in their original range, given that similar prey items and habitat are available for exploitation, with little if any interaction with native species in the Hudson Bay system.
- Temperate European and North American aquarium species have potential to colonize the Hudson Bay drainage because they are likely to be able to tolerate the prolonged winter in Canada.
- Exotic species from other continents or distant North American drainage basins have not evolved with species of Central North America and will have to create their own niche in Central North American waters. This presents a higher probability for negative interactions with North American native species.
- The carp and the rainbow smelt are probably the most destructive of the species introduced to the Hudson Bay system to date. Carp eat aquatic vegetation and are believed to increase turbidity of water by foraging in the substrate for invertebrate prey. Rainbow smelt add a new zooplanktivore and perhaps a piscivore to lakes, may lower the value of commercial fisheries by causing an 'off-flavor' in game species, and increase the number of trophic levels in ecosystems, adding to contaminant burdens of their predators.
- Public education, restriction of allowed bait fish and aquarium species, and more careful inspection of both when they are shipped across drainage boundaries are the only feasible ways, short of complete bans, to try to control transport and release of bait and aquarium fish. It must be remembered that not just fish are transferred. At the least, transferred species will carry their associated biota of multicellular parasites, microbes, and viruses, not all of which may be present in the recipient area. In addition, the water transferred with the fish will carry invertebrates, plants, etc. from the source area.

References

Atton, F.M. 1959. Invasion of Manitoba and Saskatchewan by carp. *Transactions of the American Fisheries Society* 88:203-205.

Campbell, K.B., A.J. Derksen, R.A. Remnant, and K.W. Stewart. 1991. First specimens of the rainbow smelt, *Osmerus mordax*, from Lake Winnipeg, Manitoba. *Canadian Field Naturalist* 105(4):568-570.

Carlton, J.T. and J.B. Geller. 1993. Ecological roulette: the global transport of nonindigenous marine organisms. *Science* 261:78-82.

Crossman, E.J. 1991. Introduced freshwater fishes: a review of the North American perspective with emphasis on Canada. *Canadian Journal of Fisheries and Aquatic Sciences* 48(1):46-57.

Douglas, M.E., P.C. Marsh, and W.L. Minckley. 1994. Indigenous fishes of western North America and the hypothesis of competitive displacement: *Meda fulgida* (Cyprinidae) as a case study. *Copeia* (1):9-19.

Dyke, S. 1989. Sakakawea smelt. *North Dakota Outdoors* 51(9):28-29.

Evans, D.O. and D.H. Loftus. 1987. Colonization of inland lakes in the Great Lakes region by rainbow smelt, *Osmerus mordax*: their freshwater niche and effects on indigenous fishes. *Canadian Journal of Fisheries and Aquatic Sciences* 44 (Suppl. 2):249-266

Evans, D.O., and P. Waring. 1987. Changes in the multispecies, winter angling fishery of Lake Simcoe, Ontario, 1961-83: invasion by rainbow smelt, *Osmerus mordax*, and the roles of intra- and interspecific interactions. *Canadian Journal of Fisheries and Aquatic Sciences* 44 (Suppl. 2):182-197.

Franzin, W.G., B.A. Barton, R.A. Remnant, D.B. Wain, and S.J. Pagel. 1994. Range extension, present and potential distribution, and possible effects of rainbow smelt (*Osmerus mordax*) in Hudson Bay drainage waters of northwestern Ontario, Manitoba, and Minnesota. *North American Journal of Fisheries Management* 14(1):65-76.

Gasaway, C.R. 1970. Changes in the fish population in Lake Francis Case in South Dakota in the first 16 years of impoundment. *Technical Papers of the Bureau of Sport Fisheries and Wildlife* 56:3-30.

Gould, W.R. 1981. First records of the rainbow smelt (*Osmeridae*), sicklefin chub (*Cyprinidae*) and white bass (*Percichthyidae*) from Montana. *Proceedings of the Montana Academy of Sciences* 40:9-10.

Hanke, G.F. 1996. A survey of the fishes of Lake Winnipeg and interactions of the introduced white bass with the native ichthyofauna of Hudson Bay drainage: with emphasis on young-of-the-year fishes in nearshore environments. Master's thesis. Winnipeg: University of Manitoba, Zoology Department.

Hanke, G.F. and K.W. Stewart. 1994. Evidence for northward dispersal of fishes in Lake Winnipeg. *Proceedings of the North Dakota Water Quality Symposium*, 133-149. Fargo: North Dakota State University, Water Resources Research Institute.

Henderson, B.A. and S.J. Nepszy. 1989. Factors affecting recruitment and mortality rates of rainbow smelt in Lake Erie, 1963-85. *Journal of Great Lakes Research* 15:357-366.

Hinks, D. 1943. *The Fishes of Manitoba.* Manitoba Department of Mines and Natural Resources, Winnipeg.

Jennings, M.R., and M.K. Saiki. 1990. Establishment of red shiner, *Notropis lutrensis*, in the San Joaquin Valley, California. *California Fish and Game* 76(1):46-57.

Jones, M.S., J.P. Goettl, Jr., and S.A. Flickinger. 1994. Changes in walleye food habits and growth following a rainbow smelt introduction. *North American Journal of Fisheries Management* 14:409-414.

King, T.L., E.G. Zimmerman, and T.L. Beitinger. 1985. Concordant variation in thermal tolerance and allozymes of the red shiner, *Notropis lutrensis*, inhabiting tailwater sections of the Brazos River, Texas. *Environmental Biology of Fishes* 13(1):49-57.

Koel, T.M. and J.J. Peterka. 1994. Distribution and dispersal of fishes in the Red River of the North basin: a progress report. *Proceedings of the North Dakota Water Quality Symposium*, 159-168. Fargo: North Dakota State University, Water Resources Research Institute.

Lee, D.S., C.R. Gilbert, C.H. Hocutt, R.E. Jenkins, D.E. McAllister, and J.R. Stauffer. 1980. Atlas of North American freshwater fishes. North Carolina Biological Survey Publication No. 1980-12.

Litvak, M.K. and N.E. Mandrak. 1993. Ecology of freshwater baitfish use in Canada and the United States. *Fisheries* 18(12):6-13.

Loftus, D.H. and P.F. Hulsman. 1986. Predation on boreal lake whitefish *Coregonus clupeaformis* and lake herring *C. artedii* by adult rainbow smelt *Osmerus mordax*. Canadian Journal of Fisheries and Aquatic Sciences 43:812-818.

Ludwig, H.R., D.R. Givers, and J.A. Leitch. 1994. Bait bucket movement of fish across basin boundaries. *Proceedings of the North Dakota Water Quality Symposium*, 150-157. Fargo: North Dakota State University, Water Resources Research Institute.

Manitoba Department of Natural Resources. 1994. A field guide to common bait fish species in Manitoba. Winnipeg.

Matthews, W.J. 1985. Distribution of midwestern fishes on multivariate environmental gradients, with emphasis on *Notropis lutrensis. The American Midland Naturalist* 113(2): 225-237.

Mayden, R.L. F.B. Cross and O.T. Gorman 1987. Distributional history of the rainbow smelt, *Osmerus mordax* (Salmoniformes: Osmeridae), in the Mississippi River Basin. Copeia 1987: 1051-1054.

McCulloch, B.R. 1994. *Dispersal of the Stonecat (Noturus flavus) in Manitoba and Its Interactions with Resident Fish Species*. Master's thesis. Winnipeg: University of Manitoba.

McCulloch, B.R. and K.W. Stewart. 1992. Habitat and diet comparisons between longnose dace and stonecat in the Little Saskatchewan River: interactions between a native and an invading species. *Proceedings North Dakota Water Quality Symposium*, 176-193. Fargo: North Dakota State University, Water Resources Research Institute.

Meronek, T.G., F.A. Copes, and D.W. Coble. 1995. A summary of bait regulations in the North Central United States. *Fisheries* 20(11):16-23.

Nelson, J.S. and M.J. Paetz. 1992. *The Fishes of Alberta*. Edmonton: University of Alberta Press.

Remnant, R.A. 1991. *An Assessment of the Potential Impact of the Rainbow Smelt on the Fishery Resources of Lake Winnipeg*. Master's thesis, Winnipeg: University of Manitoba.

Remnant, R.A., P.G. Graveline, and R.L. Bretecher. 1997. Range extension of rainbow smelt, *Osmerus mordax*, in Hudson Bay drainage waters of Manitoba. *Canadian Field Naturalist* 111:660-662.

Rinne, J.N. 1991. Habitat use by spikedace, *Meda fulgida* (Pisces: Cyprinidae) in southwestern streams with reference to probable habitat competition by red shiners, *Notropis lutrensis* (Pisces: Cyprinidae). *Southwestern Naturalist* 36(1):7-13.

Scott, W.B. and E.J. Crossman. 1973. Freshwater fishes of Canada. Fisheries Research Board of Canada Bulletin 184, p. 966.

Stewart, K.W. 1988. First collections of the weed shiner, *Notropis texanus*, in Canada. *Canadian Field Naturalist* 102:657-660.

Stewart, K.W. and C.C. Lindsey. 1970. First specimens of the stonecat, *Noturus flavus*, from the Hudson Bay drainage. *Journal of the Fisheries Research Board of Canada* 27(1):170-172.

Stewart, K.W. and B.R. McCulloch. 1990. The stonecat, *Noturus flavus*, in the Red River/ Assiniboine River watershed: progress report on the study of an invading species. *Proceedings of the North Dakota Water Quality Symposium*, 68-85. Fargo: North Dakota State University, Water Resources Research Insitute.

Stewart, K.W., I.M. Suthers, and K. Leavesley. 1985. New fish distribution records in Manitoba and the role of a man-made interconnection between two drainages as an avenue of dispersal. *Canadian Field Naturalist* 99(3):317-326.

Suttkus, R.D. and J.V. Connor. 1979. The rainbow smelt, *Osmerus mordax*, in the lower Mississippi River near St. Francisville, Louisiana. *American Midland Naturalist* 104: 394.

Van den Broeck, J. 1995. 1994 index netting results for Namakan, Sandpoint and Lac La Croix lakes. Ontario Ministry of Natural Resources, Fort Frances District Report. Vii + 63p.

Wain, D.B. 1993. The effects of introduced rainbow smelt (*Osmerus mordax*) on the indigenous pelagic fish community of an oligotrophic lake. M.S. thesis. Winnipeg: University of Manitoba, Zoology Department.

Walburg, C.H. 1977. Lake Francis Case, a Missouri River reservoir: changes in the fish population in 1954-75, and suggestions for management. *Technical Papers of the U. S. Fish and Wildlife Service* 95:1-12.

Wiland, L. 1995. Zebra mussel larvae found in two new Wisconsin lakes. *Littoral Drift.* July 1995. Wisconsin Sea Grant Institute.

Wright, D.G. and W.G. Franzin. 1998 (in press). The Garrison Diversion and the interbasin biota transfer issue. In *The Biology and Impacts of Some Fresh Water Invading Species in North America.* eds. Claudi, R., and J. Leach.

Zimmerman, E.G. and M.C. Richmond. 1981. Increased heterozygosity at the Mdh-B locus in fish inhabiting a rapidly fluctuating thermal environment. *Transactions of the American Fisheries Society* 110:410-416.

CHAPTER 8
Parasites and pathogens of fishes in the Hudson Bay drainage

Terry A. Dick, A. Choudhury, and B. Souter

Executive summary

Transfer of parasites and pathogens along waterways and across watersheds is a complex biological process. Initial efforts to determine potential problems are usually accomplished by comparing lists of parasites and pathogens. These lists are often incomplete because funding agencies have been reluctant to support studies relating to biodiversity. The usual solution is to compare incomplete lists; if a potentially important parasite is present in both watersheds, then the particular species of parasite or pathogen is considered to be unimportant. Studies on biodiversity, however, are key in making meaningful predictions and recommendations on potential problems resulting from transfer of water from watersheds with quite different species compositions. Pathogens such as viruses and bacteria once transferred to a new host or aquatic system are amplified directly. On the other hand, parasite communities are the end result of all aspects of interactions within an ecosystem. Factors such as changing mean annual temperatures, water quality, nutrient load and food availability affect the parasite community. Most fish parasites populations are maintained by complex interactions among the invertebrate and vertebrate hosts, trophic status of the hosts, and the nature of the food web in a given system. Change, such as the introduction of a new fish or invertebrate host can alter the parasite community and potentially affect the host population. For example, in the USSR in the 1950s, the gill parasite *Nitzschia sturionis* was introduced with a sturgeon species *Acipenser nudiventris* into the Aral Sea where a large native population of *A. nudiventris* was already present. The heavy infections of *Nitzschia* decimated the local populations of *A. nudiventris*.

One of the environmental concerns, still unanswered following the International Joint Commission report in 1977, was the transfer of fish disease agents and parasites from the Missouri River into the Manitoba waters of the Hudson Bay drainage. The Biology committee, International Garrison Diversion Study Board (1976), noted that Infectious Hematopoietic Necrosis Virus (IHNV), enteric redmouth bacterium, and *Polypodium hydriforme* might negatively impact salmonids, whitefish, and sturgeons in Lake Winnipeg. A literature search by Holloway (1983) revealed ERM to have limited negative influence on Canadian fisheries. Nevertheless, since there were sufficient concerns, research was undertaken to determine its distribution in both the Missouri and the Hudson Bay drainages. However, this report deals primarily with the Canadian situation, with specific reference to Manitoba waters.

Our studies on biota transfer concentrated on the following aspects:

- the distribution of the microorganisms, *Yersinia ruckeri,* and infectious pancreatic necrosois virus in salmonids;
- the parasites of acipenserids, especially the cnidarian parasite *Polypodium hydriforme,* which is known to infect eggs of the acipenserids; and
- an update of the previous parasites of fish species in Canadian waters.
- presenting some insights into the issues facing biota transfer as it relates to pathogens and parasites.

Microorganisms in the Hudson Bay drainage

Problem pathogens include the viruses viral hemorrhagic septicemia (Egtved virus or VHSV), infectious hematopoietic necrosis virus (IHNV), and infectious pancreatic necrosis virus (IPNV). Other pathogens important in Canada include the bacteria *Renibacterium salmoninarum* (kidney disease bacterium), *Yersinia ruckeri* (causing enteric redmouth disease), *Aeromonas samonicida* (furunculosis bacteria), and the protozoan *Myxosoma cerebralis* (causing whirling disease).

Following is a brief description of each of IHNV and *Y. ruckeri*, diseases considered important from a Canadian perspective and this biota transfer project. IHNV is an acute viral disease of fry and fingerlings of *Oncorhynchus nerka*, *O. tschawytscha*, and *O. mykiss* (Pacific drainage salmonids); but it can also be transmitted through eggs and feed. In North America, it has been reported in the United States and British Columbia. According to Schäperclaus (1992), the clinical symptoms include dark coloration, exophthalmus, abdominal distension, and petechial and hemorrhagic bleeding along the back and on the fins and pale gills. The body cavity, stomach, and intestine contain a white to yellow fluid; and the kidney, liver, and heart are pale. Necrosis of the hematopoietic tissues of the kidney, pancreas, and liver are of diagnostic value. For a complete review of this pathogen, see Schäperclaus (1992).

Yersinia ruckeri, (also called enteric redmouth, Hagerman redmouth disease, and redmouth disease has been reported from *O. mykiss* and *Salvelinus fontinalis* (see review by Schäperclaus 1992). It is a rodshaped, flagellated, gram, and oxidase negative bacterium. There are different strains but the type strain is ATCC 29473. The bacterium can be as virulent as *Aeromonas salmonicida* for fry of *Salmo salar*. The disease is characterized by increasing sluggishness; dark coloration of the fish; redness of the mouth, on the opercula, on the isthmus, and on the base of the fins. Other symptoms include hemorrhaging of the adipose tissue and hind gut, a stomach with a watery and colorless fluid, and intestine with a yellowish liquid. Fish also show acute septicemia and inflammation in most body tissues and erosion of the jaw and buccal cavity roof. It is widespread in North America, occurring in 19 states in the United States and in two provinces in Canada. Some of its transfer in North America may be by diseased fish (see discussion concerning this study). For a more detailed review of this organism, the reader is referred to Schäperclaus (1992).

The methods used in the Canadian part of the study are those set out by the Fish Health Section of the Department of Fisheries and Oceans (Fish Health Protection Regulations 1984). Surveys of wild fish populations had reported *Y. ruckeri* type II (non-pathogenic strain) in the upper Missouri River system but not from Lake Winnipeg. H. Holloway's laboratory, aware of an earlier report of *Y. ruckeri* (non-pathogenic strain) from Montana, surveyed intensively for the pathogen, but was unable to demonstrate its presence in North Dakota waters. However, *Y. ruckeri* is reported from hatcheries in Canada and is the Hagerman or type I (pathogenic) strain (Table 8.1). *Y. ruckeri* was reported from rainbow trout and brook trout in Canada (Table 8.1) and was likely brought in by diseased carrier fish to holding or hatchery areas. Its absence from wild fish suggests it is extremely rare in the Hudson Bay drainage (2 reports over 21 years) and, to date, does not appear to be established in wild fish populations.

It is also apparent from the extensive monitoring in Canada (table 8.1) that the IPN virus and bacterial kidney disease are prevalent and well established in both natural and hatchery conditions in central Canada. The fact that IHNV was not found in this study nor has it been reported over the period of 1976-1997 likely indicates that it is not present in the Hudson Bay drainage system to date.

Parasites of lake sturgeon

The cnidarian parasite *Polypodium hydriforme* was not reported from the upper Missouri River system nor the Hudson Bay drainage by Lubinsky and Loch (1979). Since it was known to infect the eggs of sturgeon species in other parts of the world and also in the United States, it was considered a potentially pathogenic parasite of sturgeon fisheries in central Canada, if transferred from the United States.

Surveys of paddlefish were undertaken on the Missouri and Yellowstone Rivers in the United States (in cooperation with Dr. H. Holloway, University of North Dakota) and by authors on lake sturgeon populations from the Nelson, Saskatchewan, Winnipeg River and Rainy Rivers in Canada. In addition, we studied sturgeon popula-

Table 8.1: Summary of the disease agents detected in the prairie region of Canada (1976-1997)*

Year	Location	Date	Agent	Species	Incidence
1976					
	WS	M	Rs	lkt	un
	WS	M	Rs	splk	un
	WS	A	Rs	bkt	1/1
	GR	Ma	IPNV	rbt&bkt	un
	WS	Ma	IPNV	rbt	3/12
	FQ	M	Rs	rbt	un
	FQ	A	Rs	bkt	un
	FQ	Ma	Rs	rbt	1/1
	FQ	Ma	Rs	bkt	1/1
	FQ	Jul	Rs	bkt	1/1
	FQ	Au	Rs	bkt	1/1
1977					
	FQ	N	Rs	rbt	3/12
	FQ	N	Rs	bkt	1/8
1978					
	FQ	D	Rs	rbt	2/24
1980					
	FQ	Jan	Rs	bkt	2/15
	FQ	Jan	Rs	rbt	6/30
	FQ	N	Rs	bkt	1/9
1981					
	RW	F	Rs	rbt	69/170
	RW	F	Rs	rbt	16/66
	RW	F	Rs	rbt	4/5
	FQ	M	Rs	rbt	8/14
	FI	Ma	Rs	rbt	2/2
	FI	Jun	Rs	rbt	2/3
	FW	Jul	Rs	rbt	8/14
	WS	S	Rs	rbt	1/14
	GR	S	IPNV	rbt	un
	GR	S	IPNV	lkt	un
	GR	S	IPNV	splk	un
	GR	S	IPNV	bkt	un
	GR	S	IPNV	bkt	un
	GR	S	IPNV	rbt	un
	GR	S	Rs	rbt	1/13
	FQ	N	Rs	rbt	2/15
	FQ	N	Rs	rbt	3/15
	RW	D	Rs	rbt	5/6
1982					
	HL	A	Rs	lkt	1/1
	RW	O	Rs	ac	1/1
	FI	O	Rs	rbt	9/15
	WS	N	Rs	bkt	4/57
	FQ	N	Rs	rbt	3/5
	FQ	N	Rs	bkt	1/5

Table 8.1: Cont.

Year	Location[a]	Date[b]	Agent[c]	Species[d]	Incidence
1983					
	WS	A	Rs	splk	2/27
	RW	Jun	Rs	rbt	3/25
	RW	Jul	Rs	rbt	2/2
	GR	S	IPNV	bkt	un
	GR	S	IPNV	splk	un
	GR	S	IPNV	rbt	un
	GR	S	IPNV	rbt	un
	GR	S	IPNV	rbt	un
	GR	S	IPNV	bkt	un
	GR	S	IPNV	bkt	un
	GR	S	IPNV	bkt	un
	GR	S	IPNV	bkt	un
	GR	S	IPNV	bkt	un
	WS	N	Rs	bkt	8/57
1884					
	WS	A	Rs	ctth	1/30
	WS	A	Rs	splk	1/30
	WS	A	Rs	lkt	7/30
	WS	A	Rs	rbt	5/30
	WS	A	Rs	ctth	5/12
	WS	A	Rs	rbt	4/12
	PBS	Ma	Y. r.	rbt	1/30
	WL	S	Rs	rbt	1/3
	CL	O	Rs	lkt	1/31
	GR	O	IPNV	bkt	un
	GR	O	Rs	lkt	4/30
	GR	O	IPNV	rbt	12/12
	GR	O	IPNV	bkt	12/12
	GR	O	Rs	bkt	3/30
	WS	O	Rs	brnt	5/30
	WS	O	Rs	rbt	5/30
	WS	O	Rs	lkt	7/30
	FQ	N	Rs	bkt	2/5
1985					
	WS	A	Rs	rbt	1/576
	WS	A	Rs	ctth	2/57
	WS	A	Rs	brnt	3/57
	GR	A	Rs	rbt	4/57
	GR	A	Rs	lkt	2/57
	GR	A	IPNV	lkt	1/11
	GR	A	Rs	lkt	2/57
	WS	O	Y.r.	bkt	1/30
	FQ	N	Rs	bkt	1/5
1986					
	GR	A	Rs	rbt	1/30
	GR	A	Rs	rbt	1/10
	GR	A	Rs	rbt	1/30
	FI	O	Rs	lkt	1/1

Table 8.1: Cont.

Year	Location	Date	Agent	Species	Incidence
1987					
	WS	M	Rs	brnt	7/30
	WS	M	Rs	rbt	7/30
	WS	M	Rs	rbt	7/30
	WS	M	Rs	rbt	16/57
	WS	M	Rs	splk	16/57
	WS	M	Rs	lkt	7/57
	WS	M	Rs	brnt	6/30
	WS	M	Rs	brnt	1/30
	WS	M	Rs	bkt	10/57
	WS	M	Rs	bkt	2/10
	WS	M	Rs	rbt	7/57
	WS	M	Rs	rbt	22/57
	RW	J	Rs	rbt	1/45
	WS	Au	Rs	rbt	1/10
	FQ	N	Rs	bkt	1/25
	FQ	N	Rs	bkt	1/20
1988					
	FQ	F	Rs	rbt	4/4
	SL	M	IPNV	rbt	9/12
	CLH	Au	IPNV	rbt	10/10
	WSW	S	Rs	lkt	1/16
1989					
	WS	S	Rs	rbt	1/50
	WS	S	Rs	rbt	1/50
	PBM	D	Rs	rbt	2/57
1990					
	WS	S	Rs	kok	7/25
	WS	S	Rs	bkt	17/50
1991					
	WS	Jan	Rs	rbt	2/50
	WS	F	Rs	bkt	1/50
	WS	F	Rs	rbt	2/50
1992					
	WS	F	Rs	rbt	1/50
	PBM	A	Rs	bkt	12/57
	WS	S	Rs	ac	26/49
1993					
	WS	M	Rs	ac	4/50
	WS	M	Rs	brnt	1/50
	WS	M	Rs	rbt	4/50
	RW	N	Rs	rbt	3/26
	RW	N	Rs	ac	2/14
1994					
	WS	M	Rs	bkt	3/25
	WS	M	Rs	bkt	7/25
1995					
	WS	M	Rs	bkt	1/10

Table 8.1: Cont.

Year	Location	Date	Agent	species	incidence
1996					
	WS	M	Rs	bkt	un
1997					
	FQ	N	Rs	rbt	un

WS= Whiteshell Hatchery, Manitoba; GR = Grand Rapids Hatchery, Manitoba; FQ = Fort Qu'appelle Hatchery; RW = Rockwood Hatchery, Manitoba; FI = Freshwater Institute, Manitoba; HL = High Lake, Manitoba; PBS =- private broker, Saskatchewan; WL = William Lake, Manitoba; SL = Sam Livingston Hatchery, Alberta; CLH = Cold Lake Hatchery, Alberta; SWS = Whiteswan Lake, Saskatchewan; PBM = private broker, Manitoba; un = unknown.

Jan = January; F = February; M= March; A = April; Ma = May; Jun = June; Jul = July, Au = August; S = September; O = October: N = November; D = December.

rbt = rainbow trout; bkt = brook trout; lake trout = lkt; splk = splake; brnt = brown trout; kok = kokanee salmon; ac = arctic char; ctth = cutthroat trout.

Rs = *Renibacterium salmonarum* (bacterial kidney disease); Y. r. = *Yersinia ruckeri* (enteric redmouth disease); IPNV = infectious pancreatic necrosis virus.

* Records of surveys, cases, and local occurrences of pathogens compiled by B. Souter.

tions in the Great Lakes drainage, specifically the Lake Winnebago/Wolf River system in the state of Wisconsin. *Polypodium hydriforme* was found in paddlefish and lake sturgeon in all the systems sampled which clearly indicates that this parasite is widely distributed in North America. Additional studies were undertaken to provide a complete list of parasites of lake sturgeon from the Hudson Bay drainage to have a detailed record of the biodiversity of the parasitofauna of this threatened species (Table 8.2). The significance of these studies concerning biodiversity is evident in the erection of two new species and the synonymizing of another species recovered from lake sturgeon. This clearly illustrates the paucity of information on lake sturgeon prior to this detailed study.

Ichthyoparasites of Manitoba: 1979-1996

This review of the parasites of Manitoba serves as an update of the 1979 report on the *Ichthyoparasites of Manitoba* by Lubinsky and Loch (1979), which compiled all available literature and information on the fish parasites of Manitoba known up until 1978 (primary, government departmental, and unpublished) and provided important baseline information for the International Garrison Diversion Unit Study Board (1976). Since then, a number of investigations into the parasites of Manitoba's freshwater fish have extended our knowledge of the fish parasites of Manitoba and are compiled in this report. The parasite fauna is compared to that reported in Lubinsky and Loch (1979) and briefly to that of North Dakota (Sutherland and Holloway 1979).

Records of parasites are from primary publications, master's and doctoral theses, government reports, from unpublished observations of the authors (TAD and AC), and from research projects done in the laboratory of Terry A. Dick. (Tables 8.3 and 8.4). The information is compiled as lists and references for each host studied between 1979-1995. Poole (1985) and Poole and Dick (1985) refer to the surveys from the Heming Lake area, and Szalai (1989) and Szalai et al. (1992) refer to the survey on Dauphin Lake fishes. Information in Watson and Dick (1979; 1980) was included by Lubinsky and Loch (1979) by reference to the master's thesis of Watson (1977) and, consequently, is not repeated in this report. New records since Lubinsky and Loch (1979) are indicated in the Host-parasite list. In table 8.3, "Heming L. area" refers to at least one of the seven boreal connected lakes investigated by Poole (1985), "gut" as site of infection refers to one or more of the different regions of the gastro-intestinal tract (stomach, caeca, intestine), and "viscera" refers to one or more visceral organs.

A total of 115 parasites (identified to species, genus, or family) has been catalogued in this report (Tables 8.3 and 8.4) from sample locations in Manitoba from 31 species of fish host belonging to 14 families.

Comparisons with the first report (Lubinsky and Loch, 1979)

Lubinsky and Loch (1979) reported the presence of 161 species of fish parasites in Manitoba. An examination of their species and source list shows that 104 of these species were exclusively from the study by Dechtiar (1972) from the Lake of the Woods. Species records in Dechtiar (1972) were included by Lubinsky and Loch (1979) since a small portion of the Lake of the Woods (mainly Buffalo Bay) falls within Manitoba boundaries and since this lake forms the headwaters of the Winnipeg River. However, the fact that 64% of the parasites listed by Lubinsky and Loch (1979) are not from within Manitoba boundaries and are from one river drainage only (Winnipeg R.) illustrates the fragmentary nature of the information on fish parasites in other waterbodies in Manitoba. Only one study (Watson 1977) provided detailed information on a parasite survey of fish species outside the Winnipeg River system.

Since the publication of Lubinsky and Loch (1979), several major surveys have added considerably to the database on fish parasites in Manitoba. These include 40 new records as well as the first detailed information from certain river drainages (Assiniboine, Red, and Nelson Rivers) and from hosts with no previous parasite records in Manitoba (e.g., *Ictalurus punctatus*, *Noturus flavus*, *Aplodinotus grunniens,* and *Morone chrysops*). In addition, 7 smelt (*Osmerus mordax*) from Lake Winnipeg were examined by us and were found to be uninfected.

Comparisons with surveys of fish parasites in North Dakota

A survey of the parasites of 24 species of fish from the James, Missouri, Sheyenne, and Wild Rice Rivers resulted in 44 species of parasites (Sutherland and Holloway 1979). Of these, only five helminths (*Icelanonchohaptor microcotyle*, *Corallotaenia minutia*, *Dacnitoides robusta*, *Neoechinorhynchus prolixus*, *Octospinifer macilentus*) and two crustaceans (*Achtheres ambloplitis* and *Ergasilus cyprinaceus*) have not been reported in Manitoba waters. However, *C. minutia*, *D. robusta,* and *A. ambloplitis* were reported from *Ictalurus melas* (*Ameiurus melas*) from the Wild Rice River and *E. cyprinaceus* was found on *Catostomus commersoni* from the Sheyenne River. Both these rivers are tributaries of the Red River, and future surveys may well find these parasites in Manitoba. Only *I. microcotyle* and *N. prolixus*, found in the Missouri River, have not been reported from the Red River drainage or other Manitoba waters. The component communities of parasites from a number of fish species (e.g., *Perca flavescens*, *Stizostedion vitreum*), particularly the helminth fauna, are similar to what has been reported from Manitoba waters.

Problems facing biota transfer studies

It would be easy to conclude that there are few remaining problems concerning biota transfer as it relates to pathogens and parasites between North Dakota waters (mainly upper Missouri River system) and the Hudson Bay drainage system. The ecosystems being impacted by potential water transfer are usually evaluated one dimensionally. For example, researchers are usually looking at the direct route of transfer rather than alternate routes. In the case of parasites and pathogens, the direct route is via the host or directly connected waterways, while the indirect route usually involves humans and natural processes. In the case of humans, this includes the movement of bait fish, enhancement or stocking programs, and aquaculture. Other forms of natural transmission include birds and mammals which may or may not follow the expected route of transferal. Any biota transfer studies that do not incorporate these issues into the overall transfer studies could lead to quite misleading interpretations and recommendations. For example, a comparision of fish parasite species lists for Dauphin Lake, Manitoba, between 1951 and 1989 found 15 new species in the system, after correcting for possible errors due to transmission of rare parasites. Six of these new parasites had direct life cycles, five were fish-transmitted and required invertebrate hosts, and four were bird-transmitted. Why this major change? The answer includes changing climatic conditions, stocking fish such as small mouth bass, increased fish eating bird populations, and possible migration of fish such as white bass into the system. We also have documented dramatic increases in the fish parasite *Contraecum* sp. in both rainbow trout and walleye (Dick et al. 1987, Dick unpublished) following stocking due to an amplification of a low level infection in the system through increased use of the system by birds and a suitable new fish host. The only change in the system was an increase in one species of fish, but both fish species after infection became unsuitable for human consumption. This tells us that changes in fish community structure can alter the transmission dynamics of a parasite and result in a relatively innocuous parasite becoming economically significant. A simple list depicting the presence or absence of a species of parasite from a watershed is a limited base on which to make meaningful predictions of what might happen.

In this case of the Garrison Diversion, we have focused on transfer to the north only. There are, however, a number of fish parasites present which are indigenous to the region and economically important to Canadian fisheries. These include the larvae of the trematodes *Apophallus brevis* (flesh and eyes), *Bolbophorus confusus* (skin and flesh), *Clinostomum complanatum* (flesh), *Hysteromorpha triloba* (flesh), and *Neochasmus umbellus* (eye); the larval nematodes *Contracaecum* spp. (viscera, body cavity, heart) and *Raphidascaris acus* (viscera, liver, intestine); larval tapeworms *Diphyllobothrium latum* (flesh) and *Triaenophorus crassus* (flesh), *Proteocephalus ambloplitis* (viscera, reproductive tract, intestine); the cnidarian *Polypodium hydriforme* (eggs); the protistans *Dermocystidium* (flesh) and *Microsporidea* (liver, flesh, viscera)

What about pathogens in Canada? The presence of *Y. ruckeri* in Canada, albeit rarely, but not in North Dakota also raises interesting questions concerning transfer. The human parasite *D. latum* is very common in the Hudson Bay drainage and infects pike, walleye, and yellow perch. Most of the other parasites are widely distributed in central North America, but their economic effects are usually expressed when water levels are low, fish eating bird populations are high, or the species mix of fish changes in a given system. In the United States, potential problems exist because considerable emphasis is placed on stocking large numbers of a single species into reservoirs for recreational purposes.

Finally, let us look at some interesting relationships (Fig.8.1). The nematode parasite *Contracecum* (a mix of at least two species) is present in Manitoba waters, and the definitive hosts are cormorants and pelicans (the populations of both are increasing on the prairies). Smelt are introduced, and its population expands dramatically in Lake Winnipeg. The

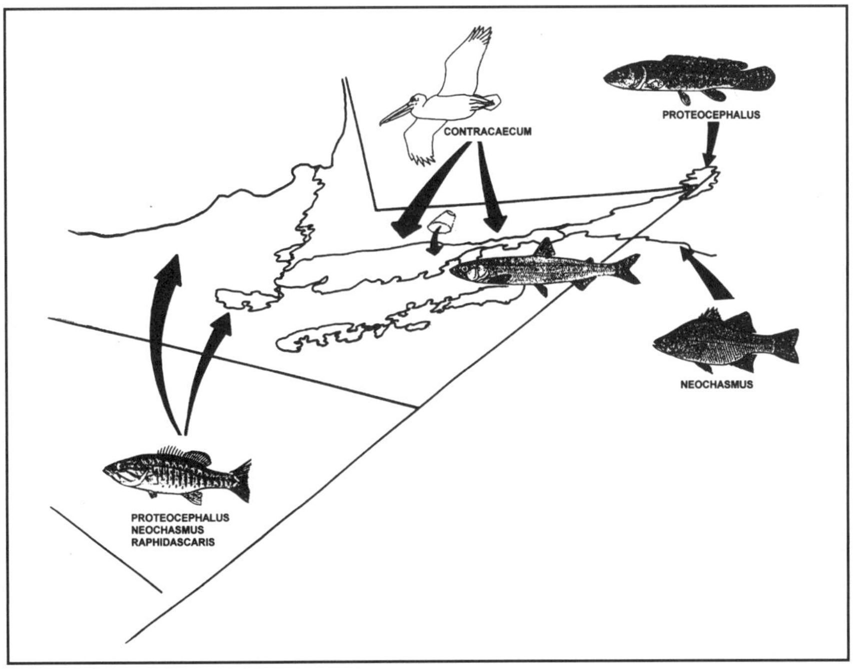

Figure 8.1 Possible routes of transfer for fish parasites.

Table 8.2: Mean abundances of infection and ranges of parasites of Lake sturgeon from different geographical locations (taken from Choudhury and Dick, 1993)

Parasite	Sask. River (N=55)* MI	Nelson River (N=131)* MI	Wpg. River (N=24)* MI	Rainy River (N=13)* MI
Monogenea				
*Diclybothrium atriatum***	0.1+0.3 (0.1) n=45	11.9+12.7 (0-86) n=141	9.7+12.9 0-46) n=19	21.7+21.6 (0-73)
Trematoda				
*Crepidostomum auriculatum***	57.0=88.0 (0-370)	383.9+447.1 (0-2484)	118.8+86.3 (0-305)	67.4+61.9 (3-208)
Diplostomum sp.(larva)	2.7+3.5 (0-14) n=45	0.5+1.5 (0-12) n=147	0.6+0.8 (0-3) n=18	0.7+1.5 (0-5)
Azygia longa	0	0	0	0.5+0.46 (0-5)
Nematoda				
*Spinitectus acipenseri***	207.8 +373.3 (0-1706)	383.9 +447.1 (0-2684)	100.1 +176.4 (0-594)	77.1+73.7 (8-244)
*Truttaed-acnitis clitellarius***	5+8.6 (0-40)	3.7+4.4 (0-27)	1.6+2.5 (0-9)	1.1+1.6 (0-5)
Raphidascaris sp.	0.2+0.6 (0-12)	0.3+1.3 (0-12)	9.0+44.1 (0-216)	1.1+1.6
Rhabdochona (*cascadilla*)	0	1.0+4.5 (0-47)	0.25+1.2 (0-6)	1.1+1.6 0
Cystidico-loides sp. (1)	0.2+0.6 (0-4)	2.5+7.2 (0-71)	0.8+1.4 (0-9)	0
Acanthocephala				
Echinorhynchus leidyi	0	0	0.6+25.1 (0-123)	0
Leptorhynchus thecatus	0	0	5.3+25.1 (0-123)	0.7+2.5 (0-9)
Neoechino-rhynchus tennulus	0	0.1+0.6 (0-4)	0	0
Pomphorhynchus bulbocolli	67.7 +140.4 (0-653)	0.9+4.0 (0-42)	1.2+5.7 (0-28)	0

Abbreviations are as follows:

MI = mean abundance of infection ± standard deviation (values in brackets are ranges of infection for that particular parasite).

Sask. River = Saskatchewan River; Wpg. River = Winnipeg River, Manitoba.

* = N numbers except when stated otherwise.

** = host specific parasites.

[1] Used as a minimum estimate for 216 cysts with a minimum of one nematode /cyst.

Table 8.3: Host-parasite list

Parasites within brackets [ ] have been reported previously in Manitoba (in Lubinsky and Loch 1979) from the host under which they appear.

CHONDROSTEI

Acipenseridae

Acipenser fulvescens (Lake sturgeon):

Diclybothrium atriatum, [*Crepidostomum lintoni*], *Cystidicolides* sp.(l), [*Truttaedacnitis clitellarius*], *Raphidascaris* sp. (l), *Rhabdochona cascadilla*, *Spinitectus acipenseri*, *Echinorhynchus leidyi*, *Leptorhynchoides thecatus*, *Neoechinorhynchus tenellus*, *Pomphorhynchus bulbocolli*, *Placobdella montifera*, *Ichthyomyzon unicuspis*. (Nelson R., Winnipeg R., Choudhury and Dick, 1993).
Polypodium hydriforme (Nelson R., Winnipeg R.; Choudhury and Dick 1993).

TELEOSTEI

Salmonidae

Coregonus clupeaformis (Lake whitefish):

Crepidostomum cooperi, [*C. farionis*], *Prosorhynchoides pusilla*, *Bothriocephalus cuspidatus*, [*Diphyllobothrium dendriticum*], [*Glaridacris catostomi*], [*Ligula intestinalis*], [*Proteocephalus exiguus*], [*Proteocephalus* sp.], [*Triaenophorus crassus*], *Raphidascaris acus* (l), *Rhabdochona milleri*, *Spinitectus gracilis*, [*Pomphorhynchus bulbocolli*], *Ergasilus versicolor* (Heming Lake area, Poole, 1985).
Also: Heterophyiidae (l), *Bolbophorus* sp. (l) (The Pas, Choudhury and Dick, unpublished).

C. artedii (Ciscoe): ([1]-Poole, 1985; [2]-Szalai, 1989)

Argulus appendiculosus [2], *Ergasilus luciopercarum* [2], *Proteocephalus wickliffi* [2], *R. acus* (l) [1,2], *Tetracotyle* sp. (l) [2], *D. dendriticum* [1], *Triaenophorus nodulosus* [2]

Oncorhynchus mykiss (Rainbow trout, aquaculture):

Contracaecum sp. (Dick 1987), *T. crassus* (Whiteshell, authors, unpublished). *Tetracotyle* sp., *Neascus* sp.

Salmo trutta (Brown trout, aquaculture):

T. crassus (Whiteshell, Dick, unpublished).

Salvelinus namaycush (Lake trout):

Cystidicola sp., *Raphidascaris* sp. (l). (Nopiming, Choudhury and Dick, unpublished).

Esocidae ([1]- Poole, 1985; [2]-Szalai, 1989)

Esox lucius (Northern pike):

[*Tetraonchus monenteron*[1,2]], [Azygia longa[2]], [*Centrovarium lobotes*[2]], Crepidostomum cooperi[1], [*Crepidostomum farionis*[1]], *Diplostomum* sp. (l)[2], Lissorchis kritskyi[1], *Neascus* sp. (l)[2], Prosorhynchoides pusilla[1], [*Bothriocephalus cuspidatus*[1,2]], [*Diphyllobothrium latum*[1]], [*Glaridacris catostomi*[1]], *Ligula intestinalis* [1], [*Proteocephalus pinguis*[1,2]], Schistocephalus solidus[1], [*Triaenophorus crassus*[1]], [*Triaenophorus nodulosus*[1,2]], *Contracaecum* sp. (l)[2], Raphidascaris acus[1,2], *Raphidascaris acus* (l)[1], Rhabdochona milleri[1], *Spinitectus gracilis* [1,2], *Tetracotyle* sp. (l) [2], *Neoechinorhynchus strigosus* / N. rutili[1], *Pomphorhynchus bulbocolli*[1], Unionidae[2], Argulus appendiculosus[2], Ergasilus luciopercarum[1]. Placobdera montifera[2].
Also: [*Azygia longa*], *Raphidascaris* sp. (l) (encapsulated on lumenal side of stomach wall (lakes in the Whiteshell area, Choudhury and Dick, unpublished).

Hiodontidae

Hiodon alosoides (Goldeye):

Azygia longa, *Crepidostomum illinoiense*, *Lissorchis crassicrurum*, *Paurorhynchus hiodontis*, *Bothriocephalus cuspidatus*, *Ligula intestinalis*, *Raphidascaris acus* (a,l), *Rhabdochona canadensis*, [*Ergasilus nerkae*] (Szalai, 1989).
Also: *Centrovarium lobotes* (Plum R.), *Cystidicoloides tenuissima*, *C. illinoiense*, *P. hiodontis* and *B. cuspidatus* (Assiniboine R., Red R.) (Choudhury and Dick, unpublished).

Hiodon tergisus (Mooneye):

Crepidostomum illinoiense, *Paurorhynchus hiodontis*, *Cystidicoloides tenuissima* (Glenn, 1980).

Table 8.3: Cont.

Cyprinidae

Cyprinus carpio (Common carp):
Capillaria sp., *Pomphorhynchus bulbocolli* (Red R., Choudhury, unpublished observations).

Notropis atherinoides (Emerald shiner):
Centrovarium lobotes (l), *Diplostomum* sp. (l), *Contracaecum* sp. (l), *Argulus appendiculosus* (Szalai 1989).

Notropis cornutus (Common shiner):
C. lobotes (l), *Diplostomum* sp. (l), *Pomphorhynchus bulbocolli*, *Spinitectus gracilis.* (Szalai 1989).

Notropis hudsonius (Spottail shiner):
C. lobotes, *Diplostomum* sp. (l), *Bothriocephalus cuspidatus*, [*Ligula intestinalis*], *Proteocephalus pinguis*, *Triaenophorus nodulosus*, *Contracaecum* sp. (l), *Raphidascaris acus* (l), *P. bulbocolli*, *Myzobdella moorei* (Szalai 1989).

Pimephales promelas (Fathead minnow):
Bolbophorus sp. (l), *Raphidascaris* sp. (l),(authors unpublished).

Catostomidae

Catostomus commersoni (White sucker) ([1]-Poole 1985; [2]-Szalai 1989):
[*Myxosoma* sp.[2]], [*Allocreadium lobatum*[1]], *Asymphylodora amnicolae*[1], [*Diplostomum* sp.(l)[1,2]], *Lissorchis crassicrurum*[2], *L. kritskyi*[1], *Neascus* sp.[1], *Prosorhynchoides pusilla*[1], [*Tetracotyle* sp.[2]], *Biacetabulum* sp.[2], [*Glaridacris catostomi*[1]], [*Hunterella nodulosa*[2]], [*Ligula intestinalis*[1,2]], *Proteocephalus* sp.[1], *Schistocephalus solidus*[1], *Contracaecum* sp.(l)[2], *Philometra nodulosa*[2], *Philonema* sp.[1], *Raphidascaris acus* (l)[1,2], *Rhabdochona canadensis*[2], *R. milleri*[1], *S. gracilis*[1], [*Neoechinorhynchus crassus*[1,2]], *N. distractus*[2], *N. rutili*[1], [*N. strigosus*[1]], [*Pomphorhynchus bulbocolli*[1,2]], Unionidae[2], *Argulus appendiculosus*[2], [*Ergasilus versicolor*[1,2]], *Lernaea cyprinacea*[2], *Myzobdella moorei*[2], *Piscicola* sp.[1], *Placobdera montifera*[2].

Moxostoma macrolepidotum and *M. anisurum* (Redhorses):
[*Myxosoma* sp.], [*Diplostomum* sp.], [*Lissorchis crassicrurum*], [*Neascus* sp.], [*Tetracotyle* sp.], *Biacetabulum infrequens*, *Biacetabulum* sp., [*Khawia iowensis*], *Contracaecum* sp.(l), [*Raphidascaris acus*] (l), [*Rhabdochona canadensis*], [*Neoechinorhynchus crassus*], [*N. cristatus*], [*N. distractus*], [*Pomphorhynchus bulbocolli*], [*Ergasilus versicolor*], *Cystobranchus verilli*, [*Myzobdella moorei*], [*Placobdella montifera*] (Szalai 1989).

Carpiodes cyprinus (Quillback)
[*Anonchohaptor anomalum*], *Myxosoma*, *Diplostomum* sp. (l), *Lissorchis gularis*, *Monobothrium hunteri*, [*Rowardleus pennensis*], *Neoechinorhynchus carpiodi*,[*Pomphorhynchus bulbocolli*], *Argulus appendiculosus Ergasilus lizae*, [Unionidae]. (Szalai 1989)

Ictaluridae

Ictalurus punctatus (Channel cat):
Alloglossidium geminum, *Crepidostomum ictaluri*, *Clinostomum* sp.(l), [*Phyllodistomum lacustri*], *Polylekithum ictaluri*, *Rhabdochona* sp., *Spinitectus* n.sp. (mature), *S. gracilis* (immature), *Corallobothrium fimbriatum*, *Megathylacoides* (*Corallobothrium*) giganteum (Red R., Choudhury and Dick, unpublished and in prep.).

Ameiurus melas (Black bullhead):
Diplostomum spathaceum, *Phyllodistomum staffordi*, *Hysteromorpha triloba* (l) (Cook's creek, Choudhury and Dick, unpublished).

Noturus flavus (Stonecat):
Crepidostomum ictaluri, *Diplostomum spathaceum*, *Phyllodistomum lacustri*, *Polylekithum ictaluri* (Assiniboine R., Red R., Choudhury and Dick, unpublished).

Gadidae

Lota lota (Burbot):
Raphidascaris sp., (Dick, unpublished), *Pomphorhynchus bulbocolli* (Glenn, 1980).

Percopsidae

Percopsis omiscomaycus (Troutperch):
Urocleidus adspectus, Centrovarium lobotes (l), *Crepidostomum cooperi*, *Diplostomum* sp. (l), *Tetracotyle* sp. (l), *Raphidascaris acus* (l), *Rhabdochona canadensis*, [*Spinitectus gracilis*], *Argulus appendiculosus*, *Myzobdella moorei*, (Szalai 1989; Szalai *et al.* 1992).

Table 8.3: Cont.

Also: *Myxosoma* sp., *Crepidostomum n. sp.* (Dauphin L., P. Nelson *et al.* 1987. *Pomphorhynchus bulbocolli*, *Ergasilus* sp., *Neochasmus* sp. (l), *Bolbophorus* sp. (l) (Dauphin L., P. Nelson and T.A. Dick, unpublished observations).

Gasterosteidae

Culaea inconstans (brook stickleback):

Contracaecum sp. (l), (High Rock L., Dick 1987; Dick *et al.* 1987).

Centrarchidae

Amblopites rupestris (Rockbass):

Spinitectus gracilis, *Rhabdochona* sp. (Red R., Choudhury, unpublished).

Micropterus dolomieui (Smallmouth bass)

Posthodiplostomum minimum, *Bothriocephalus cuspidatus*, *Contracaecum* sp. (l), *Pomphorhynchus bulbocolli*, *Ergasilus luciopercarum* (Szalai, 1989).

Also: *Clinostomum* sp. (l), *Diplostomum* sp. (l), *Proteocephalus ambloplitis*, *Raphidascaris acus*, (Wellman L., Choudhury, and Dick, unpublished observations; Whiteshell lakes, Dick and Choudhury, unpublished).

Bucephalidae, *Crepidostomum cornutum*, *Cryptogonimus chyli*, *Microphallus* sp., *Spinitectus gracilis*, *S. carolini*, *Leptorhynchoides thecatus* (Winnipeg River, Choudhury, unpublished).

Moronidae

Morone chrysops (White bass):

Urocleidus chrysops, *Neochasmus umbellus*, *Bothriocephalus cuspidatus* (Imm), *Camallanus oxycephalus*, *Rhabdochona* sp. (L. Winnipeg, P. Nelson, unpublished, Choudhury and Dick, unpublished observations).

Sciaenidae

Aplodinotus grunniens (Freshwater drum):

Pomphorhynchus bulbocolli (Szalai 1989; Szalai *et al.* 1992)

Microcotyle spinicirrus, *Crepidostomum auritum*, *Homalometron* sp., *Rhabdochona* sp. (Red R., Choudhury, unpublished observations).

Percidae ([1]-Poole 1985; [2]-Szalai, 1989)

Perca flavescens (Yellow perch):

[*Urocleidus adspectus*[2]], [*Bunodera sacculata*[1]], [*Apophallus brevis*[1]], *Azygia longa*[2], *Caecincola* sp.[2], [*Centrovarium lobotes*[2]], [*Crepidostomum cooperi*[1,2]], *C. farionis*[1], *Creptotrema funduli*[2], [*Diplostomum* sp.[2]], *Lissorchis kritskyi*[1], *Ornithodiplostomum ptychocheilus*[1], *Prosorhynchoides pusilla*[1], *Rhipidocotyle papillosus*[1], *Tetracotyle* sp.(l)[2], [*Bothriocephalus cuspidatus* (l)[1,2]], [*Diphyllobothrium latum*[1]], *Proteocephalus pearsei*[1,2], *Proteocephalus ambloplitis* (l); *Schistocephalus solidus*[1], [*Triaenophorus nodulosus* (l)[1,2]], *Contracaecum* sp. (l)[2], [*Raphidascaris acus* (l)[1,2]], R. *acus*[2], *Rhabdochona milleri*[1], *Neoechinorhynchus rutili*[1], *P. bulbocolli*[1], [*Spinitectus gracilis*[1,2]], [*Ergasilus luciopercarum*[1,2]], *Myzobdella moorei*[2], *Placobdella montifera*[2], Unionidae[2].

Also: [*Dichelyne cotylophora*] (Whiteshell lakes, Choudhury and Dick, unpublished), [*Crepidostomum cooperi*] (Round L., Pigeon R., Dick and Choudhury, unpublished).

Stizostedion canadense (Sauger):

Crepidostomum cooperi, [*Centrovarium lobotes*], *Creptotrema funduli*, [*Bothriocephalus cuspidatus*], [*Proteocephalus luciopercae*], *Raphidascaris acus*. (Szalai 1989).

Stizostedion vitreum vitreum (Walleye):

[*Urocleidus aculeatus*[1]], [*U. adspectus*[2]], [*Bunodera sacculata*[1]], [*Centrovarium lobotes*[2]], *Crepidostomum cooperi*[1,2], *Creptotrema funduli*[2], *Diplostomum* sp. (l)[1,2], *Neascus* sp.[2], *Ornithodiplostomum ptychocheilus* (l)[1], *Prosorhynhcoides pusilla*[1], *Rhipidocotyle papillosus*[1], [*Bothriocephalus cuspidatus*[1,2]], [*Diphyllobothrium latum*[1]], *Ligula intestinalis* (l)[1], [*Proteocephalus luciopercae*[1,2]], *Schistocephalus solidus* (l)[1], *T. nodulosus*[1], *Spinitectus gracilis*[1], [*Raphidascaris acus*[2]], R. *acus* (l)[2], *Rhabdochona milleri*[1,2], [*Neoechinorhynchus crassus*[1]], *N. strigosus*[1], [*N. tenellus*[2]], [*Pomphorhynchus bulbocolli*[1]], *Argulus appendiculosus*[2], [*Ergasilus luciopercarum*[2]], *Myzobdella moorei*[2], *Placobdella montifera*[2], Unionidae[2].

Percina caprodes (Logperch):

Centrovarium lobotes, [*Diplostomum* sp.[2]], *Raphidascaris acus* (l), *Spinitectus gracilis*, *Pomphorhynchus bulbocolli*, *Myzobdella moorei* (Szalai 1989).

Table 8.4: Parasite list with their hosts and waterbodies in which they occur

Parasite	Host	Site	Water body
MYXOSPORIDEA			
Myxosoma	*Catostomus commersoni*	Gills	Dauphin L.
	Percopsis omiscomaycus	Muscle	Dauphin L.
CNIDARIA			
*Polypodium hydriforme**	*Acipenser fulvescens*	Eggs	Nelson R., Winnipeg R.
PLATYHELMINTHES			
Trematoda			
Allocreadium lobatum	*Catostomus commersoni*	Gut	Heming L. area
Apophallus brevis (l)	*Perca flavecsens*	Muscle	Heming L. area
*Asymphylodora amnicolae**	*Catostomus commersoni*	Gut	Heming L. area
Azygia longa	*Esox lucius*	Stomach	Dauphin L., Whiteshell Lakes
	Hiodon alosoides	Gut	Dauphin L.
	Perca flavescens	Stomach	Dauphin L., Whiteshell Lakes
Bolbophorus sp.(l)*	*Coregonus clupeaformis*	Subcutaneous	The Pas district lakes
	Pimephales promelas	Subcutaneous	Pothole ponds
	Percopsis omiscomaycus	Flesh	Dauphin L.
Bucephalidae	*Micropterus dolomieui*	Gut	Winnipeg R.
Bunodera sacculata	*Perca flavescens*	Gut	Heming L. area
	Stizostedion vitreum	Gut	Heming L. area
Caecincola sp.	*Perca flavecsens*	Gut	Dauphin L.
Centrovarium lobotes (a)	*Esox lucius*	Gut	Dauphin L.
	Hiodon alosoides	Gut	Plum R.
	Perca flavescens	Gut	Dauphin L.
	Stizostedion canadense	Gut	Dauphin L.
	Stizostedion vitreum	Gut	Heming L. area
	Percina caprodes	Gut	Dauphin L.
Centrovarium lobotes (l)	*Notropis atherinoides*	Muscle	Dauphin L.
	Notropis cornutus	Muscle	Dauphin L.
	Notropis hudsonius	Muscle	Dauphin L.
	Percopsis omiscomaycus	Muscle	Dauphin L.
Clinostomum sp. (l)	*Ictalurus punctatus*	Subcutaneous	Red R.
	Micropterus dolomieui	Subcutaneous	Wellman L.
Crepidostomum auriculatum	*Acipenser fulvescens*	Gut	Nelson R., Winnipeg R.
*C. auritum**	*Aplodinotus grunniens*	Gut	Red R.
C. cooperi	*Coregonus clupeaformis*	Gut	Heming L. area
	Esox lucius	Gut	Heming L. area
	Percopsis omiscomaycus	Gut	Dauphin L.
	Perca flavescens	Gut	Dauphin L., Heming L. area, Pigeon R.
	Stizostedion canadense	Gut	Dauphin L.
	Stizostedion vitreum	Gut	Dauphin L., Heming L. area
C. cornutum	*Micropterus dolomieui*	Gut	Winnipeg R.
C. farionis	*Perca flavescens*	Gut	Heming L. area
	Coregonus clupeaformis	Gut	Heming L. area

Table 8.4: Cont.

Parasite	Host	Site	Water body
	Esox lucius	Gut	Heming L. area
C. ictaluri	*Ictalurus punctatus*	Gut	Red R.
	Noturus flavus	Gut	Red R., Assiniboine R.
*C. illinoiense**	*Hiodon alosoides*	Gut	Dauphin L., Red R., Assiniboine R., Winnipeg R.
	Hiodon tergisus	Gut	Assiniboine R.
Crepidostomum n. sp.*	*Percopsis omiscomaycus*	Gut	Dauphin L.
*Creptotrema funduli**	*Perca flavescens*	Gut	Heming L. area
	Stizostedion canadense	Gut	Dauphin L.
	Stizostedion vitreum	Gut	Dauphin L.
Cryptogonimus chyli	*Micropterus dolomieui*	Gut	Winnipeg R.
Diplostomum spathaceum (l)	*Ameiurus melas*	Eyes (lens)	Cook's Creek
	Noturus flavus	Eyes (lens)	Assiniboine R.
Diplostomum sp. (l)	*Esox lucius*	Eyes	Dauphin L.
	Notropis atherinoides	Eyes	Dauphin L.
	Notropis cornutus	Eyes	Dauphin L.
	Notropis hudsonius	Eyes	Dauphin L.
	Catostomus commersoni	Eyes	Dauphin L.
	Moxostoma macrolepidotum	Eyes	Dauphin L., Heming L. area
	Carpiodes cyprinus	Eyes	Dauphin L.
	Percopsis omiscomaycus	Eyes	Dauphin L.
	Micropterus dolomieui	Eyes	Wellman L.
	Perca flavescens	Eyes	Dauphin L.
	Stizostedion vitreum	Eyes	Heming L. area, Dauphin L.
	Percina caprodes	Eyes	Dauphin L.
Heterophyiidae (l)	*Coregonus clupeaformis*	Muscle	The Pas district lakes
*Hysteromorpha triloba**	*Ameiurus melas*	Muscle	Red R., Cook's Creek
*Homalometron (armatum?)**	*Aplodinotus grunniens*	Gut	Red R.
*Lissorchis crassicrurum**	*Hiodon alosoides*	Gut	Dauphin L.
	Catostomus commersoni	Gut	Dauphin L.
	Moxostoma macrolepidotum	Gut	Dauphin L.
*Lissorchis gularis**	*Carpiodes cyprinus*	Gut	Dauphin L.
*Lissorchis kritskyi**	*Esox lucius*	Gut	Heming L. area
	Catostomus commersoni	Gut	Heming L. area
	Perca flavescens	Gut	Heming L. area
Microphallus sp.	*Micropterus dolomieui*	Stomach	Winnipeg R.
Neascus sp. (l)	*Coregonus clupeaformis*	Skin	The Pas
	Esox lucius	Skin	Dauphin L.
	Moxostoma macrolepidotum	Skin	Dauphin L.
	Catostomus commersoni	Gut	Heming L. area
	Stizostedion vitreum	Gut	Dauphin L.
Neochasmus umbellus (a)*	*Morone chrysops*	Gut	L. Winnipeg (South basin)
Neochasmus umbellus (l)*	*Percopsis omiscomaycus*	Eyes	Dauphin L.
Ornithodiplostomum ptychocheilus (l)	*Perca flavescens*	Viscera	Heming L. area
	Stizostedion vitreum	Viscera	Heming L. area

Table 8.4: Cont.

Parasite	Host	Site	Water body
Paurorhynchus hiodontis	*Hiodon alosoides*	Body cavity, gut	Dauphin L., Red R., Assinniboine R.
	Hiodon tergisus	Body cavity, gut	Assiniboine R.
*Phyllodistomum lacustri**	*Ictalurus punctatus*	Urinary bladder	Red R.
	Noturus flavus	Urinary bladder	Red R., Assiniboine R.
Phyllodistomum staffordi	*Ameiurus melas*	Urinary bladder	Red R.
Polylekithum ictaluri	*Ictalurus punctatus*	Gut	Red R.
	Noturus flavus	Gut	Red R., Assiniboine R.
Posthodiplostomum minimum (l)	*Micropterus dolomieui*	Liver, spleen	Dauphin L.
Prosorhynchoides pusilla	*Coregonus clupeaformis*	Gut	Heming L. area
	Esox lucius	Gut	Heming L. area
	Catostomus commersoni	Gut	Heming L. area
	Perca flavescens	Gut	Heming L. area
	Stizostedion vitreum	Gut	Heming L. area
*Rhipidocotyle papillosus**	*Perca flavescens*	Gut	Heming L. area
	Stizostedion vitreum	Gut	Heming L. area
Tetracotyle sp.	*Catostomus commersoni*	Pericardium	Dauphin L.
	Moxostoma macrolepidotum	Pericardium	Dauphin L.
	Perca flavescens	Pericardium	Dauphin L.
	Coregonus artedii	Pericardium	Dauphin L.
	Oncorhynchus mykiss	Pericardium	Whiteshell hatchery
	Percopsis omiscomaycus	Pericardium	Dauphin L.
	Esox lucius	Pericardium	Dauphin L.
Monogenea			
*Diclybothrium atriatum**	*Acipenser fulvescens*	Gills	Nelson R., Winnipeg R.
*Microcotyle spinicirrus**	*Aplodinotus grunniens*	Gills	Red R.
Anonchohaptor anomalum	*Carpiodes cyprinus*	Ureters	Dauphin L.
Tetraonchus monenteron	*Esox lucius*	Gills	Dauphin L., Heming L. area
Urocleidus aculeatus	*Stizostedion vitreum*	Gills	Heming L. area
U. adspectus	*Perca flavescens*	Gills	Dauphin L., Heming L. area
	Percopsis omiscomaycus	Gills	Dauphin L.
	Stizostedion vitreum	Gills	Dauphin L.
*U. chrysops**	*Morone chrysops*	Gills	L. Winnipeg
Cestoda			
Biacetabulum infrequens	*Moxostoma macrolepidotum*	Gut	Dauphin L.
	M. anisurum	Gut	Dauphin L.
Biacetabulum sp.	*Catostomus commersoni*	Gut	Dauphin L.
Bothriocephalus cuspidatus	*Coregonus clupeaformis*	Gut	Heming L. area
	Esox lucius	Gut	Heming L. area
	Hiodon alosoides	Gut	Dauphin L., Red R., Assiniboine R.
	H. tergisus	Gut	Assiniboine R.
	Notropis hudsonius	Gut	Dauphin L.
	Micropterus dolomieui	Gut	Dauphin L.
	Perca flavescens	Gut	Dauphin L., Heming L. area, Whiteshell area
	Stizostedion canadense	Gut	Dauphin L.

Table 8.4: Cont.

Parasite	Host	Site	Water body
	S. vitreum	Gut	Dauphin L., Heming L. area
B. cuspidatus (Immature)	*Morone chrysops*	Gut	L. Winnipeg (South basin)
Corallobothrium fimbriatum	*Ictalurus punctatus*	Gut	Red R., Assiniboine R.
Diphyllobothrium latum (l)	*Esox lucius*	Muscle	Heming L. area, Whiteshell lakes
	Perca flavescens	Muscle	Heming L. area, Whiteshell lakes
	Stizostedion vitreum	Muscle	Heming L. area
D. dendriticum (l)*	*Coregonus clupeaformis*	Viscera	Heming L. area
	C. artedii	Viscera	Heming L. area
*Eubothrium rugosum**	*Lota lota*	Gut	Whiteshell area
Eubothrium sp.	*Salvelinus namaycush*	Gut	Unnamed lake in Nopiming Park
Glaridacris catostomi	*Coregonus clupeaformis*Gut		Heming L. area
	Esox lucius	Gut	Heming L. area
	Catostomus commersoni	Gut	Heming L. area
Hunterella nodulosa	*Catostomus commersoni*	Gut	Dauphin L.
*Khawia iowensis**	*Moxostoma macrolepidotum*	Gut	Dauphin L.
	Moxostoma anisurum	Gut	Dauphin L.
Ligula intestinalis (l)	*Coregonus clupeaformis*	Body cavity	Heming L. area
	Esox lucius	Body cavity	Heming L. area
	Hiodon alosoides	Body cavity	Dauphin L.
	Notropis hudsonius	Body cavity	Dauphin L.
	Catostomus commersoni	Body cavity	Dauphin L., Heming L. area
	Stizostedion vitreum	Body cavity	Heming L. area
*Megathylacoides giganteum**	*Ictalurus punctatus*	Gut	Red R.
Monobothrium hunteri	*Carpiodes cyprinus*	Gut	Dauphin L.
Proteocephalus ambloplitis (a)	*Micropterus dolomieui*	Gut	Wellman L.
P. ambloplitis (l)	*Micropterus dolomieui*	Viscera, ovaries	Wellman L.
	Perca flavescens	Viscera	Whiteshell area
P. exiguus	*Coregonus clupeaformis*	Gut	Heming L. area
P. luciopercae	*Stizostedion vitreum*	Gut	Dauphin L., Heming L. area
	S. canadense	Gut	Dauphin L.
P. pearsei	*Perca flavescens*	Gut	Dauphin L., Heming L. area
P. pinguis	*Notropis hudsonius*	Gut	Dauphin L.
	Esox lucius	Gut	Dauphin L., Heming L. area
*P. wickliffi**	*Coregonus artedii*	Gut	Dauphin L.
Proteocephalus sp.	*Coregonus clupeaformis*	Gut	Heming L. area
	Catostomus commersoni	Gut	Heming L. area
*Rowardleus pennensis**	*Carpiodes cyprinus*	Gut	Dauphin L.
Schistocephalus solidus	*Esox lucius*	Body cavity	Heming L. area
	Catostomus commersoni	Body cavity	Heming L. area
	Perca flavescens	Body cavity	Heming L. area
	Stizostedion vitreum	Body cavity	Heming L. area

Table 8.4: Cont.

Parasite	Host	Site	Water body
Triaenophorus crassus	*Esox lucius*	Gut	Heming L. area
T. crassus (l)	*Coregonus clupeaformis*	Muscle	Heming L. area
	Oncorhynchus mykiss	Muscle	Whiteshell hatchery
	Salmo trutta	Muscle	Whiteshell hatchery
T. nodulosus	*Esox lucius*	Gut	Dauphin L., Heming L. area
T. nodulosus (l)	*Coregonus artedii*	Viscera	Dauphin L.
	Notropis hudsonius	Viscera	Dauphin L.
	Perca flavescens	Liver	Dauphin L., Heming L. area
NEMATODA			
Camallanus oxycephalus	*Morone chrysops*	Gut	L. Winnipeg (South basin)
Capillaridae	*Cyprinus carpio*	Gut	Red R.
Contracaecum sp. (l)*	*Oncorhynchus mykiss*	Viscera, muscle	High Rock L.
	Esox lucius	Viscera	Dauphin L.
	Notropis atherinoides	Viscera	Dauphin L.
	Pimephales promelas	Viscera	High Rock L.
	Catostomus commersoni	Viscera	Dauphin L.
	Moxostoma macrolepidotum	Viscera	Dauphin L.
	M. anisurum	Viscera	Dauphin L.
	Culaea inconstans	Heart (atrium)	High Rock L.
	Micropterus dolomieui	Viscera	Dauphin L.
	Perca flavescens	Viscera	Dauphin L.
Cystidicola (*farionis* ?)	*Salvelinus namaycush*	Swim bladder	Unnamed lake in Nopiming Park
*Cystidicoloides tenuissima**	*Hiodon alosoides*	Gut	Red R., Assiniboine R.
	H. tergisus	Gut	Assiniboine R.
Cystidicoloides sp. (imm.)	*Acipenser fulvescens*	Gut	Nelson R.
Dichelyne cotylophora	*Perca flavescens*	Gut	Whiteshell Lakes
*Philometra nodulosa**	*Catostomus commersoni*	Subcutaneous	Dauphin L.
Philonema sp.	*Catostomus commersoni*	?	Heming L. area
Raphidascaris acus	*Esox lucius*	Gut	Dauphin L., Heming L. area
	Hiodon alosoides	Gut	Dauphin L.
	Micropterus dolomieui	Gut	Wellman L.
	Stizostedion vitreum	Gut	Dauphin L.
R. acus (l)	*Coregonus clupeaformis*	Gut wall	Heming L. area
	C. artedii	Viscera	Dauphin L., Heming L. area
	Esox lucius	Viscera, stomach wall	Heming L. area, Whiteshell lakes
	Hiodon alosoides	Viscera	Dauphin L.
	Notropis hudsonius	Viscera	Dauphin L.
	Pimephales promelas	Liver	Whiteshell area
	Catostomus commersoni	Gut, viscera	Dauphin L., Heming L. area
	Moxostoma macrolepidotum	Viscera	Dauphin L.
	M. anisurum	Viscera	Dauphin L.
	Lota lota	Liver	?

Table 8.4: Cont.

Parasite	Host	Site	Water body
	Percopsis omiscomaycus	Liver	Dauphin L.
	Perca flavescens	Liver	Dauphin L., Heming L. area
	Stizostedion canadense	Liver	Dauphin L.
	S. vitreum	Liver	Dauphin L., Heming L. area
	Percina caprodes	Liver	?
Raphidascaris sp. (l)	*Acipenser fulvescens*	Forestomach	Winnipeg R.
	Salvelinus namaycush	Forestomach	Unnamed lake in Nopiming Park
*Rhabdochona canadensis**	*Hiodon alosoides*	Gut	Dauphin L.
	Catostomus commersoni	Gut	Dauphin L.
	Moxostoma (2 spp.)	Gut	Dauphin L.
	Percopsis omiscomaycus	Gut	Dauphin L.
R. cascadilla (imm.)	*Acipenser fulvescens*	Gut	Nelson R., Winnipeg R.
*R. milleri**	*Coregonus clupeaformis*	Gut	Heming L. area
	Esox lucius	Gut	Heming L. area
	Catostomus commersoni	Gut	Heming L. area
	Perca flavescens	Gut	Heming L. area
	Stizostedion vitreum	Gut	Heming L. area
Rhabdochona sp.	*Ictalurus punctatus*	Gut	Red R.
	Ambloplites rupestris	Gut	Red R.
	Morone chrysops	Gut	L. Winnipeg (South basin)
	Aplodinotus grunniens	Gut	Red R.
*Spinitectus acipenseri**	*Acipenser fulvescens*	Stomach	Nelson R., Winnipeg R.
S. carolini	*Micropterus dolomieui*	Stomach	Winnipeg R.
S. gracilis	*Coregonus clupeaformis*	Stomach	Heming L. area, Winnipeg R.
	Esox lucius	Gut	Dauphin L., Heming L. area
	Notropis cornutus	Gut	Dauphin L.
	Catostomus commersoni	Gut	Heming L. area
	Ambloplites rupestris	Gut	Red R.
	Micropterus dolomieui	Gut	Winnipeg R.
	Perca flavescens	Gut	Dauphin L., Heming L. area
	Percina caprodes	Gut	Dauphin L.
	Stizostedion vitreum	Gut	Heming L. area
S. gracilis (Immature)	*Ictalurus punctatus*	Gut	Red R.
Spinitectus n. sp.*	*Ictalurus punctatus*	Gut	Red R., Assiniboine R.
*Truttaedacnitis clitellarius**	*Acipenser fulvescens*	Gut	Nelson R., Winnipeg R.
ACANTHOCEPHALA			
*Echinorhynchus leidyi**	*Acipenser fulvescens*	Gut	Winnipeg R.
Leptorhynchoides thecatus	*Micropterus dolomieui*	Gut	Winnipeg R.
L. thecatus (imm.)	*Acipenser fulvescens*	Gut	Winnipeg R.
*Neoechinorhynchus carpiodi**	*Carpiodes cyprinus*	Gut	Dauphin L.
N. crassus	*Catostomus commersoni*	Gut	Dauphin L., Heming L. area
	Moxostoma macrolepidotum	Gut	Dauphin L.

Table 8.4: Cont.

Parasite	Host	Site	Water body
	M. anisurum	Gut	Dauphin L.
	Stizostedion vitreum	Gut	Heming L. area
N. cristatus	*Moxostoma macrolepidotum*	Gut	Dauphin L.
	M. anisurum	Gut	Dauphin L.
*N. distractus**	*Catostomus commersoni*	Gut	Dauphin L.
	Moxostoma macrolepidotum	Gut	Dauphin L.
	M. anisurum	Gut	Dauphin L.
N. rutili	*Esox lucius*	Gut	Heming L. area
	Catostomus commersoni	Gut	Heming L. area
	Perca flavescens	Gut	Heming L. area
N. strigosus	*Esox lucius*	Gut	Heming L. area
	Catostomus commersoni	Gut	Heming L. area
	Stizostedion vitreum	Gut	Heming L. area
N. tenellus	*Stizostedion vitreum*	Gut	Dauphin L.
N. tenellus (Imm.)	*Acipenser fulvescens*	Gut	Nelson R.
Pomphorhynchus bulbocolli	*Coregonus clupeaformis*	Gut	Heming L. area
	Esox lucius	Gut	Dauphin L., Heming L. area
	Hiodon tergisus	Gut	Assiniboine R.
	Cyprinus carpio	Gut	Red R.
	Notropis cornutus	Gut	Dauphin L.
	N. hudsonius	Gut	Dauphin L.
	Catostomus commersoni	Gut	Dauphin L., Heming L. area
	Moxostoma macrolepidotum	Gut	Dauphin L.
	M. anisurum	Gut	Dauphin L.
	Carpiodes cyprinus	Gut	Dauphin L.
	Lota lota	Gut	Assiniboine R.
	Percopsis omiscomaycus	Gut	Dauphin L.
	Micropterus dolomieui	Gut	Dauphin L.
	Aplodinotus grunniens	Gut	Dauphin L.
	Perca flavescens	Gut	Heming L. area
	Percina caprodes	Gut	Dauphin L.
	Stizostedion vitreum	Gut	Heming L. area
P. bulbocolli (Immature)	*Acipenser fulvescens*	Gut	Nelson R.
CRUSTACEA			
Argulus appendiculosus	*Coregonus artedii*	Body surface	Dauphin L.
	Esox lucius	Body surface	Dauphin L.
	Notropis hudsonius	Body surface	Dauphin L.
	Catostomus commersoni	Body surface	Dauphin L.
	Carpiodes cyprinus	Body surface	Dauphin L.
	Percopsis omiscomaycus	Body surface	Dauphin L.
	Stizostedion vitreum	Body surface	Dauphin L.
Ergasilus luciopercarum	*Coregonus artedii*	Gills	Dauphin L.
	Micropterus dolomieui	Gills	Dauphin L.
	Perca flavescens	Gills	Dauphin L., Heming L. area
	Stizostedion vitreum	Gills	Dauphin L.
*E. lizae**	*Carpiodes cyprinus*	Gills	Dauphin L.
	Esox lucius	Gills	Heming L. area

Table 8.4: Cont.

Parasite	Host	Site	Water body
E. versicolor	*Coregonus clupeaformis*	Gills	Heming L. area
	Catostomus commersoni	Gills	Dauphin L., Heming L. area
	Moxostoma macrolepidotum	Gills	Dauphin L.
	M. anisurum	Gills	Dauphin L.
Ergasilus sp.	*Percopsis omiscomaycus*	Gills	Dauphin L.
*Lernaea cyprinacea**	*Catostomus commersoni*	Skin	Dauphin L.
ANNELIDA			
Hirudinea			
*Cystobranchus verilli**	*Moxostoma macrolepidotum*	Body surface	Dauphin L.
Myzobdella moorei	*Notropis hudsonius*	Body surface	Dauphin L.
	Catostomus commersoni	Body surface	Dauphin L.
	Moxostoma macrolepidotum	Body surface	Dauphin L.
	M. anisurum	Body surface	Dauphin L.
	Percopsis omiscomaycus	Body surface	Dauphin L.
	Perca flavescens	Body surface	Dauphin L.
	Percina caprodes	Body surface	Dauphin L.
	Stizostedion vitreum	Body surface	Dauphin L.
Piscicola sp.	*Catostomus commersoni*	Body surface	Heming L. area
*Placobdella montifera**	*Acipenser fulvescens*	Body surface	Winnipeg R.
	Esox lucius	Body surface	Dauphin L.
	Catostomus commersoni	Body surface	Dauphin L.
	Moxostoma macrolepidotum	Body surface	Dauphin L.
	M. anisurum	Body surface	Dauphin L.
	Perca flavescens	Body surface	Dauphin L.
	Stizostedion vitreum	Body surface	Dauphin L.
MOLLUSCA			
Unionidae (l)	*Esox lucius*	Gills	Dauphin L.
	Catostomus commersoni	Gills	Dauphin L.
	Moxostoma macrolepidotum	Gills	Dauphin L.
	Moxostoma anisurum	Gills	Dauphin L.
	Carpiodes cyprinus	Gills	Dauphin L.
	Perca flavecens	Gills	Dauphin L.
	Stizostedion vitreum	Gills	Dauphin L.
CHORDATA			
*Ichthyomyzon unicuspis**	*Acipenser fulvescens*	Body surface	Winnipeg R.

* New records for Manitoba. ? = Site unknown. g = glochidia.

non-host specific *Contracecum* larvae increase in abundance in the smelt. We know that walleye can be infected with *Contracecum* larvae (Dick, unpublished), walleye can eat smelt, and *Contracecum* can be transferred laterally in the food chain. For example, piscivores such as rainbow trout or walleye can acquire the parasite from a smaller fish (Dick et. al. 1987, Dick, unpublished). Obviously the presence of the parasite, albeit at very low levels now, in the smelt, walleye, and fish eating birds is all important.

The second example is the parasitic tapeworm, *P. ambloplitis*, reported in the Hudson Bay drainage only in lakes stocked with infected bass *Micropterus dolomieui* in small, isolated lakes in the Whiteshell; the larvae have been found in yellow perch from some of these lakes (Dick unpublished). This parasite infects the viscera and ovaries of fish and is reported to cause parasitic castration. We also know that the fish *Amia calva*, although apparently absent from Manitoba, harbors *P. ambloplitis*. If the parasite should enter the yellow perch fishery in Lake Winnipeg or elsewhere, who is responsible? After all, it is one of the parasites species listed for the Hudson Bay drainage. This discussion was not developed to assess blame, but rather to point out that the problems of pathogens and parasite transfer are multidimensional and that their impact may be cumulative over an extended time period and depend on factors in addition to the simple presence of pathogen and parasites and, as a consequence may be much more difficult to predict than previously believed. However, retrospective studies on complex aquatic communities will enable scientists and policy makers to determine cause and effect if good baseline data and adequate monitoring over time are available on the key components of the impacted ecosystem.

Summary

We are able to conclude the following:

- *Y. ruckeri* in Canada is rarely present in fish populations, but appears to be restricted to the stocking or hatchery industries at the moment;
- to date, IHNV has not been isolated, indicating that it is not present in the Hudson Bay drainage or North Dakota waters;
- both *Y. ruckeri* and IHNV remain potential problems to the Hudson Bay drainage if water is transferred, accidently or purposely, without prior treatment to kill these microrganisms (this statement is based on proof that both pathogens are present in the Missouri River system);
- the cnidarian parasite *P. hydriforme* is present in lake sturgeon in the Hudson Bay drainage and in paddlefish in the Missouri River system. It is a normal part of the parasitofauna of lake sturgeon and paddlefish;
- the discovery of two new species of parasites from lake sturgeon indicates the lack of information on biodiversity available at the beginning of this study;
- the 40 new records of parasite in fish from Canadian waters also indicates the lack of information at the start of this study and the importance of biodiversity studies;
- while "new records" may indicate the transfer of pathogens and parasites by waterways, there are other routes of entry such as importation by the aquaculture/hatchery industry, movement of bait fish and other aspects of the recreation industry, and increased abundance of birds carrying potential pathogens or parasites that affect the commercial value of fish;
- the movement and increase in numbers of fish hosts such as the white bass can affect the numbers of parasites in a system;
- the changing mix of fish hosts harboring parasites which impact on the commercial fishing industry and/or stocking may not be related to biota transfer but to other factors such as global warming.

This means that future studies dealing with biota transfer must be multidisciplinary, highly integrated concerning the collection of data, focused on a few key indicator species, and must have the vision to predict "natural processes" which may otherwise confound biota transfer studies as background noise. Good baseline biodiversity data are critical and if they are insufficient to make predictions, then its collection must have the highest priority in the initial stages of a study relating to biota transfer.

Acknowledgements

The authors acknowledge the financial support of the North Dakota Water Resources Research Institute Interbasin Biota Transfer program. We would also like to thank the many fishermen and biologists in both the United States and Canada that helped make this study possible. T. Dick also thanks graduate students, T. deVos and A. Szalai, for helping with collections of paddlefish eggs and serum from the Missouri and Yellowstone Rivers. T. Dick is particularly grateful for the timing of this funding since it allowed the collection and compilation of biological data from lake sturgeon populations from the last commercial fisheries in western Canada. Lake sturgeon is a heritage species in Manitoba and a threatened species in much of its range in Canada. Biological samples of lake sturgeon are no longer available due to the closing of all commercial fisheries in Western Canada. All sturgeon species worldwide are now listed under CITES. T. Dick also acknowledges the financial support of the Science Subvention programme of the Department of Fisheries and Oceans, Government of Canada and the Natural Sciences and Engineering Research Council of Canada.

References

International Garrison Diversion Unit Study Board. 1976. Appendix C. Biology Report.

Choudhury, A. and T.A. Dick. 1992. *Spinitectus acipenseri* n. sp. (Nematoda: Cystidicolidae) from the lake sturgeon Acipenser fulvescens Rafinesque in Canada. *Systematic Parasitology* 22: 131-140.

Choudhury, A. and T.A. Dick. 1993. Parasites of the lake sturgeon, *Acipenser fulvescens* Rafinesque, 1817 (Chondrostei: Acipenseridae) from Central Canada. *Journal of Fish Biology* 42(4): 571-584.

Dechtiar, A.O. 1972. Parasites of fish from Lake of the Woods, Ontario. *Journal of the Fisheries Research Board of Canada* 29: 285-293.

Dick, T.A. 1987. The atrium of the fish heart as a site for *Contracaecum* spp. larvae. *Journal of Wildlife Diseases* 23: 328-330.

Dick, T.A., H. Holloway, and A. Choudhury. 1991. Polypodium sp. (Coelentenrate) from lake sturgeon (Acpinser fulvescens Rafinesque) from the prairie region of Canada. J. Parasitology 77: 483-484.

Dick, T.A. and R. Rosen. 1981. Identification of *Diplostomum* spp. from the eyes of lake whitefish, *Coregonus clupeaformis* (Mitchill), based on experimental infection of herring gull chick, Larus argentatus Pontoppidan. *Canadian Journal of Zoology* 59(6): 1176-1179.

Dick, T.A., M.H. Papst, and H.C. Paul. 1987. Rainbow trout (*Salmo gairdneri*) stocking and *Contracaecum* spp. *Journal of Wildlife Diseases* 23: 242-247.

Fish Health Protection Regulations Manual of Compliance. 1984. Miscellaneous special publication 31 (revised), Department of Fisheries and Oceans, Fisheries Research Directorate, Ottawa, Canada. pp. 39.

Holloway, H.L., Jr. 1983. An update on a fish parasite and diseases considered by the International Joint Commission in the transbasin diversion of Missouri River water in to the Hudson Bay drainage. pp. 37-67. In: Clambey, G.K., H.L. Holloway, Jr., J.Owen, and J.J. Peterka. *Potential Transfer of Aquatic Biota Between Drainage Systems Have no Natural Flow Connection.* 76 pp.

Lubinsky, G.A. and J.S. Loch. 1979. Ichthyoparasites of Manitoba: literature review and bibliography. *Canadian Fisheries and Marine Service Manuscript Report* 513: iv + 29 p.

Nelson, P.A., A. Choudhury, and T. Dick. 1997. Crepidostomum percopsisi n. sp. (Digenea: Allocreadiidoc) from trout perch (Percopsis omiscomaycus) of Dauphin Lake, Manitoba. J. Parasitology 83: 1157-1160.

Poole, B.C. 1985. *Fish-parasite Population Dynamics in seven Small Boreal Lakes of Central Canada.* M.S. thesis, University of Manitoba, Winnipeg.

Poole, B.C. and T.A. Dick. 1985. Parasite recruitment by stocked walleye, *Stizostedion vitreum* vitreum (Mitchill), fry in a small boreal lake in Manitoba. *Journal of Wildlife Diseases* 21(4): 371-376.

Szalai, A.J. 1989. *Factors Affecting Community Structure, Transmission and Regulation of Fish Parasites in Dauphin Lake, Manitoba.* Ph.D. thesis, University of Manitoba, Winnipeg.

Schaperclaus, W. 1992. *Fish Diseases.* Vols. 1, 2, A.A. Balkema, Rotterdam.

Sutherland, D.R. and H.L. Holloway. 1979. Parasites of fish from the Missouri, James, Sheyenne, and Wild Rice Rivers in North Dakota. *Proceedings of the Helminthological Society of Washington* 46(1): 128-134.

Szalai, A.J., J.F. Craig and T.A. Dick. 1992. Parasites of fishes from Dauphin Lake, Manitoba 1985-1987. *Canadian Technical Report of Fisheries and Aquatic Sciences*, 1735.

Watson, R.A. 1977. *Metazoan Parasites from Whitefish, Cisco, and Pike from Southern Indian Lake, Manitoba. A Pre-empoundment and Diversion Analysis.* MSc. thesis, University of Manitoba, Winnipeg.

CHAPTER 9
Water Treatment

Don Richard, Robert A. Zimmerman, Karl E. Rosvold, and G. Padmanabhan

Executive summary

Interbasin transfer of water raises questions of introduction of foreign biota across watershed boundaries. The engineering aspects of transfer of huge quantities of water have been addressed sufficiently in the past to the extent that many such transfer schemes are in operation all over the world. However, applications of water treatment technologies to control biota transfer have not been addressed adequately in the past Environmental engineers need to adapt traditional water treatment systems, or develop new ones to reduce the risk of biota transfer. Furthermore, these tasks need to be accomplished within economic, environmental, and social constraints. This chapter reviews studies on water treatment methods such as screening, filtration, and disinfection in relation to interbasin water transfer, particularly Garrison Diversion transfer in North Dakota. In the Garrison Diversion project, a fish screen facility was investigated in the past for removing fish, fish eggs, and fish larvae. Fish pathogen removal was found infeasible. Direct filtration through sand filters was also investigated. A pilot scale facility was constructed and tested. Also investigated were some methods of disinfection such as chlorination, ozonation, and ultraviolet exposure. Huge magnitudes of water volumes in these types of transfers continue to pose special problems of treatment.

Introduction

All diversions of water across watershed boundaries require engineering expertise. Traditionally interbasin water transfers have required canals, dams, pipelines, pumping stations, drop structures, and associated appurtenances. The design of such structures and providing the necessary control equipment have typically been the responsibility of water resources engineers. Environmental engineers came aboard to address concerns related to public health and environmental quality. Ecosystem changes, due to interbasin water transfers, include changes in water quality, quantity, habitat alteration, and introduction of nonindigenous aquatic species (Padmanabhan et al. 1990). Although plenty of literature is available on interbasin water transfers, relatively little information is available on the transfer of biota resulting from these projects (Padmanabhan et al. 1990; Jensen 1991). This is especially true when considering using water treatment technologies to control biota transfer. The need to prevent the foreign biota introduction has expanded the environmental engineer's role in interbasin water transfers to include applying water treatment technologies to situations typically dominated by water resources engineering problems. If the goal of zero biota transfer is to be attained, water resources engineers must design conveyance systems, reservoirs, and other hydraulic structures that will prevent the transfer of untreated water across the watershed boundary even in severe circumstances such as flooding or impending structural failure. Environmental engineers must adapt traditional, or develop new, water treatment systems to design a foolproof system which will adequately reduce the risk of biota transfer. These are formidable tasks that also must be accomplished under economic, environmental, and social constraints. Engineering aspects of preventing biota transfer through water transfer are a significant challenge. Research into the application of water

treatment techniques for preventing biota transfer in the Garrison Diversion Project, North Dakota, U.S.A., is the focus of this chapter.
Specifically, three different water treatment technologies have been included in research related to biota transfer in the Garrison Diversion Unit:

- Screening
- Direct filtration
- Disinfection

Screening

The first method aimed at preventing the transfer of biota via the Garrison Diversion Unit was a fish screen facility. The design first proposed called for all the water in the McClusky Canal to be passed through metal screens designed to remove fish, fish eggs, and fish larvae (Johnson 1975). Construction of the facility was stopped in 1977 when it became apparent that the screen would not remove viruses and bacteria responsible for fish diseases and that some fish eggs and larvae may pass into the Hudson Bay drainage (Turner 1988). There was also considerable concern over the operation of the facility, especially when dealing with emergency situations. It was clear that as designed, the screens could break, or clog, leading to the inadvertent transfer of untreated water (International Garrison Diversion Study Board 1976). The concerns over the inadequacies of the designed fish screen facility led to a detailed evaluation of the efficacy of the McClusky Canal Fish Screen Facility and, eventually, a new design (U.S. Bureau of Reclamation 1982). Although concerns over operational integrity and the ability of the screens to remove viruses and bacteria from the water still exist, the U.S. Bureau of Reclamation (USBR) had demonstrated that the technology is feasible to remove fish, fish eggs, and larvae.

In the design of the first facility, the USBR considered several methods, such as poisons, violent hydraulic action, electrocution, ultrasonics, ozone, and extreme pressures, to control the movement of all fish present in the Upper Missouri River that would pose problems in the Hudson Bay watershed. Based on harmful effects on other wildlife and on suspect reliability, all options except screening and filtering were rejected, and finally, screening was selected as the more economical alternative. Despite the huge scale, the Bureau's Engineering and Research Center designed and constructed a full size model and a full-scale facility based on model testing (U.S. Department of the Interior 1982, Johnson 1975).

Water would flow onto fixed horizontal (or at a slight dip) metal screens. Underneath the top screen, a coarse screen would be provided to help support the finer screen. The screens would be woven from fine metal strands and have a fabric-like texture that would allow the screens to be largely self-cleaning. The metal screens would trap material larger that the openings on the screen allowing the water and smaller particles to pass. On a clean screen, the water would impact close to the side of the screen where the water was entering and quickly pass through the screen. As the screen slowly clogged with debris, the water would spread out toward the far end of the screen, moving the trapped debris. When the debris reached the far side of the screen, it would be washed into a trough or some sort of collection system and be removed for disposal.

The durable metal strands in the screen, which may be about 0.2 millimeters wide (0.008 inches), are woven so that the openings are very small. The 1975 design called for using a 40-mesh metal screen. The openings in a 40-mesh screen are 0.432 millimeters (0.017 inches) (Turner 1988). It was obvious, that if the fish pathogens as well as the fish, fish eggs, and larvae themselves were to be screened, these openings were too large. Bacteria and viruses can be smaller than 0.001millimeters (0.00004 inches). Furthermore, the fish eggs and larvae of some of the species of fish considered a potential threat were smaller than the openings in a 40-mesh screen. The fish eggs and larvae of concern can be smaller than 1 millimeter.

In addition to these physically small but potentially dangerous biota passing the screen, there were several significant design and operational concerns that the proposed fish screen would be ineffective. Many of these concerns were addressed by the International Joint Commission (International Joint Commission 1977). The U.S. Bureau of Reclamation made significant improvements over the 1975 design. After extensive research in Bureau laboratories, a small screening facility near Turtle Lake, North Dakota, was constructed to test their new design (U. S. Department of the Interior 1982).

It was apparent that a fish screen that could stop the pathogens in the water may not be feasible. Water passing through any screen is restricted by metal strands that comprise the screen. The smaller the openings, the less water that can pass through the screen without pressurizing the flow. For example, the 40-mesh screen has a nominal open area of about 46 percent while a 70-mesh screen (used in the 1982 design) has an open area of only 30 percent. Therefore, there is an inherent trade off between how quickly the water can be passed through a screening facility and how small of a particle can be effectively screened. Considering the size of viruses, it is questionable whether even tightly woven textiles could pass water by gravity and still remove the viruses. If it were feasible, the size of the screens would have to be immense to pass the quantity of water even in a scaled back project. Pressurizing the flow might have been an option but would have added to pumping costs and required a new design for holding the screen or fabric in place while guaranteeing that the fabric would not be torn or punctured. This was not feasible or economical. Therefore, it was concluded the goal of the new design should not be to prevent all biota transfer, but to remove the fish eggs and larvae.

Screen size was the first issue to be addressed in the new design. Tests were conducted at the Engineering & Research Center of the U.S. Bureau of Reclamation in Denver, Colorado, and the Tennessee Valley Authority (TVA) laboratories in Norris, Tennessee, to scientifically establish that the new screen design would be adequate to remove all fish eggs and larvae. To determine the opening size requirements, the task force reviewed the list of fish species cited by the International Garrison Diversion Study Board (1976) as potentially problematic and selected four species for detailed analysis. The species selected were the rainbow smelt, Utah chub, carp, and gizzard shad.

The sizes of fresh fish eggs and larvae were measured using innovative photographic techniques. The smallest eggs were from rainbow smelt and gizzard shad, and the smallest larvae were from the Utah chub and gizzard shad. Originally, a 60-mesh screen had been recommended for the new design. The 60-mesh screen has openings of 0.234 millimeters (0.0092 centimeters). The 70-mesh screen recommended in the final 1982 design called for openings of 0.198 millimeters (0.008 inches). The 60- and 70-mesh screens, therefore, appeared to be small enough to capture the larvae and eggs tested. Along with a few other screens, these types were tested in laboratory experiments.

Another concern about the screens was the ability to manufacture screens to design specifications and to monitor the quality. The actual measurements on a 70-mesh screen showed that, although the normal opening size was about 0.2 millimeters, the openings varied from 0.17-0.27 millimeters (0.0067-.0107 inches). Notably, the smallest fish eggs and larvae were larger than the largest measured openings.

The review of the 1975 design also called for demonstrating the integrity of the design (International Garrison Diversion Study Board 1976). Some novel experiments involved testing different types and sizes of screens in a standpipe apparatus. Actual fish eggs and larvae, some of which were dyed, were added to the top of a vertical pipe. Water flowed down the pipe; at the bottom, the water passed through the test screen and a fine net used to capture plankton from lake water. No whole larvae (only severely damaged pieces of larvae) passed through the 60-mesh screen, but several whole smelt larvae did pass through the 40-mesh screen and were captured in the plankton net. This confirmed that the 1975 design calling for a 40-mesh screen was inadequate.

Another critical concern about the screens included the durability of the material. Thus, the screens used at the Brekken-Holmes test facility near Turtle Lake, North Dakota, were studied. To satisfy the requirements of the International Garrison Diversion Study Board, the Bureau built a small fish screen facility within the Missouri River basin. The screens were evaluated during field tests conducted over eight months in two years.

A photographic procedure was used to verify that the opening sizes did not change with extended use, and to provide the opportunity to visually inspect the screens for signs of wear or damage. Results indicated the monel screen opening sizes did not change. A phosphor-bronze material was also tested at Brekken-Holmes. The performances of the monel and phosphor-bronze screens were comparable when the screens were new; however, the hydraulic capacity of the phosphor-bronze screens deteriorated rapidly with use. At least some of the deterioration was attributed to abrasion. Photographs revealed that the phosphor-bronze screens corroded, filling the openings with deteriorating metal. The monel screens did not exhibit these problems.

Another concern was damage that could be caused by material puncturing the screen or individual strands breaking. The photographs showed that normal operation did not cause any breaks. However, screens were damaged during handling. Because it was impractical to photograph every screen in its entirety, an easy way to detect damage was needed. In most instances, if there was damage, especially a puncture, it would be visually apparent due to the fine texture of the screen, but a more efficient method was developed. The screen was submerged in water, and air was introduced a few centimeters below the screen. The air would emerge first from any damaged areas, or even larger openings, due to imperfect manufacturing methods. This proved to be an effective method of determining the integrity of the screens.

After testing with the standpipe apparatus, sectional models were constructed and tested at the Engineering and Research Center in Denver and at the Hydraulic Laboratory at Norris. This extended the evaluation to include how the screens were fitted into their frames and how the frames would be sealed into place. If any eggs or larvae could pass between the screen and the frame, or pass the seals, the integrity of the facility would also be compromised.

Tests with the different size screens and the old seal design indicated that no matter what screen size was tested some whole larvae passed through the section model. The model seal was based on compression of a rubber gasket that was used in the 1975 design. The seals were siliconed to try to improve their performance, but occasional seal failures did occur, resulting in the passage of complete larvae. This led to the development of rectangular, innertube-like, pneumatic seals that proved effective. The rectangular tubes had to be fabricated in the laboratory because they were not commercially available, but the effort was rewarded with excellent operational performance in laboratory tests and at the Brekken-Holmes facility. The seals were fitted around the screen frames and inflated with air to maintain a pressurized seal around the screens, preventing water from passing around the screens. This design had two distinct advantages over the previous design. First, if the seal was damaged, the air pressure would begin to drop within the air line that maintains the pressure within the innertube. Therefore, the pressure can be monitored, and any loss in pressure can signal the operators to shut off that set of screens and locate the seal deficiency. This type of operational safeguard was suggested as an indispensable part of any screen used on the Garrison Diversion Project.

Water impacting the screen, an uneven buildup of debris on the screens, or a sudden loading of debris may cause the screens to shift. The old seal design of a gasket compressed by bolts was relatively rigid and any shift in position would likely have compromised the screen integrity. During testing of the pneumatic seal design at the Brekken-Holmes facility distortions of over 15 centimeters (up to six inches) in the screen frames were observed without any indications of seal failure. The frames were later reinforced, further limiting the possibility of seal failure.

Part of the efficiency of horizontal screens is their self-cleaning characteristics. As the screen collects more debris, water moves toward the downstream end of the screen where debris is carried over the screen. However, the screens occasionally still needed mechanical cleaning. Several types of water-spray cleaning designs were investigated. Based on testing at the Brekken-Holmes facility, a traveling rack of nozzles was suggested for the final design. Each screen is equipped with three spray bars driven back and forth across the length of the screen by air pressure and pulleys. Each bar consists of nozzles to spray water over the screen and remove debris. Two of the bars are situated between the main screen and the lower backup screen. One of these bars cleans the top screen from below, and the other cleans the bottom screen from above. The third bar cleans the bottom screen from below. Results indicated that this system was adequate to remove the accumulated debris. During several occasions, the screens at the Brekken-Holmes facility clogged, or were allowed to clog, to a point where several inches of water sat stagnant over the screen. Even under these conditions, the traveling spray nozzle bar was sufficient to clean the screens.

In the final design of the modified McClusky Canal Fish Screen Facility (U.S. Bureau of Reclamation 1982; U.S. Department of the Interior 1982) water would pass through trashracks with the flow being controlled by radial gates immediately behind the trashracks. The openings on the trashrack would be 3.8 centimeters (1.5 inches). Similar types of screens are commonly found on intakes to municipal and industrial water treatment plants using rivers, lakes, or reservoirs for water supply. The objective of the trashracks is to remove large material from the influent water. Typically, the trashracks would be able to remove fish, large plants, branches, rocks, plastic, and other large material from the water. Debris clinging to the trashracks would be removed by a rake that travels up and down the trashrack. The debris collected by the rake would be fed to a conveyor for disposal.

The water passing the trashracks would then enter the fish screen bays. Debris retained on the screens would be removed by natural cleaning action and spray washing and overflow into debris troughs. Then the water would flow to a canal or reservoir and eventually to the Hudson Bay watershed. The debris and water collected by the fish screens would be stored in a debris sump and then pumped to another screening structure for dewatering.

Any debris collected by the facility presents a disposal problem. Whether the debris is stored in a lagoon or hauled to a landfill, any water collected with the debris takes up space and adds weight, reducing the storage capacity of lagoons and increasing the cost of landfill disposal. The design calls for 80-mesh vibrating screens that should pass most of the water and collect debris even smaller than the 70-mesh screens used in the main structure. Water passing the vibrating screens would be pumped back to the canal on the influent side of the facility. The design calls for debris to be stored in lagoons.

The modifications and new features added to the 1975 design were expected to effectively prevent movement of fish, fish eggs, and fish larvae cited as problematic species in the Garrison Diversion Project. Besides the physical features, detailed monitoring and operational guidelines were outlined to strictly limit potential failures, even under extreme circumstances. However, significant doubts remained. To prevent transfer of all aquatic species of concern (i.e., bacteria, viruses, and protozoans), screened water would have to be disinfected, and possibly both filtered and be disinfected. For example, the protozoan *Polypodium hydriforme* is larger than the openings on the proposed screen; but this parasite's consistency resembles jelly and, therefore, may also pass through the screens (Clambey et al. 1983). Furthermore, there are concerns about the ability of a single fish screen structure to meet the zero biota transfer requirements sought in the Garrison Diversion project. An alternative is to have multiple barriers at different locations within the Garrison Diversion Unit (Clambey et al. 1983; Sayler 1990). After all, there are possible mechanisms of biota transfer other than through water transfer (Ludwig and Leitch 1995). Based on the economics of constructing, maintaining and operating multiple facilities of this magnitude, it is believed that more complete treatment of Garrison Diversion water may be more practical than providing multiple fish screen facilities (Sayler 1990).

Direct filtration

Passing water through sand filters was one of the original preventive measures considered by the U.S. Bureau of Reclamation in the Garrison Diversion Project (U.S. Department of the Interior 1982). At the time, it was clear that a fish screen would be economical, compared to sand filtration (Harza Engineering 1977) and if the screen would satisfy international concerns, filtration should be avoided. However, it also became clear that international concerns were not limited to fish but other biota; specifically the bacterium *Yersinia ruckeri*, the Infectious Hematopoietic Necrosis Virus (IHNV), and the protozoan *Polypodium hydriforme*. Because fish screens could not reliably remove these threats to wild and farmed fish populations in the Hudson Bay watershed, direct filtration was investigated as an alternative.

In drinking water treatment, sand filters are typically preceded by coagulation, flocculation, softening, and sedimentation processes. As opposed to traditional filtration, direct filtration is not preceded by any softening or sedimentation processes. The important distinction is that in traditional filtration, the filters remove material that does not readily settle by gravity in settling basins, while in direct filtration, the filters must remove both settleable and nonsettleable solids. Therefore, for water to be suitable for direct filtration, it must be low in suspended solids (Turner 1988). Garrison Diversion water is low in suspended solids and in the types of algae that tend to clog sand filters (Hefta 1990).

Inflow usually passes through, or over, troughs to a pool of water that rests on the filter media. The filter media usually consist of a layer of sand beneath a layer of coal. The water flows by gravity through the filter media where solids are captured within the media. Filtered water is collected in an underdrain system, usually coarse gravel, and routed to a storage basin called the clear well. From the clear well, the water can then be disinfected, if not disinfected already, and pumped to consumers.

When clogged, the filters are cleaned by a process called backwashing. During backwashing, the flow of water entering the filter is stopped, and filtered water from the clear well is pumped back through the underdrain system into the filter media. Enough water is pumped back to fluidize the bed. Fluidization forces the individual grains of media apart. The fluidization of the filter media, caused by the rapid back washing flow of water, allows material trapped in the media to be removed. The backwash water, along with the material removed from the filter media, flows into a wash trough and is routed to other treatment processes or disposal. Because the filters must be cleaned or backwashed, it is necessary to supply more than one filter in any large water treatment plant.

As water flows by gravity through the filter media, solids are removed by at least four different processes. Straining occurs when solids are trapped between at least two adjacent particles of the media. If a particle impacts a piece of media, the media is said to intercept the solid. In some cases, solids within the media actually settle by gravity onto a media particle. This is termed sedimentation. Finally, some very small particles, or even dissolved particles, may undergo a surface reaction phenomenon and become chemically bonded to the filter media or solids already captured by the media. As water continues to flow through the filter bed, more of the solids that are trapped in the upper layers may break free and become captured lower in the filter media. In this manner, filtration can continue until the entire depth of the filter is accumulating solids.

As solids accumulate in the filter, the resistance to water flow increases because there are fewer open spaces for the water to flow. Clogged media tend to decrease the flow of water through the filter, which in turn leads to water backing up above the filter media, and a pressure drop, or head loss, in the media. When the filter clogs, as indicated by one of the above factors, the media are backwashed. Filter runs may also have to be ended due to solids breakthrough. If the water passing through the filter starts carrying solids through the media, the effective depth has been used up, and the filter must be backwashed, even if the resistance to flow is not excessive.

When dual media (coal-sand) filters are used, a larger depth of media can be used to capture more solids before the filter requires backwashing. Filtered water must be used to backwash the filters. Because dual media filters store more solids and result in a lower resistance to the flow of water, the filter runs are longer than filters with a single layer of sand media. Therefore, only a smaller portion of the finished water is required to backwash dual media filters. When the U.S. Bureau of Reclamation initially evaluated filtration during the development of fish screens, this type of dual media filter was uncommon and relatively unproven. The lower flow rates and greater backwashing required in a single sand layer configuration made filtration uneconomical compared to the fish screen. With the development of the dual media configuration and knowledge of the process variables, dual media direct filtration became a viable alternative to fish screens.

Because coal particles are larger than sand particles, only the large solids in the influent water are captured in the coal layer, allowing the smaller particles to pass into the sand layer where they can be trapped. In this manner, the entire depth of the filter media can be used to store solids without resulting in an excessive resistance to flow (head loss). Therefore, more water can be filtered before the media clog and require backwashing. If the entire bed consisted of sand, the solids would rapidly clog the upper surfaces of the filter, resulting in rapid increases in head loss and frequent backwashing. This phenomenon is referred to as blinding the filter.

Direct filtration tests were performed to design appropriate media configuration and evaluate process variables to see if direct filtration was an applicable treatment process for treating Garrison Diversion water (Turner and Hefta 1989; Hefta 1990; Moretti et al. 1993). In this research, it was confirmed that viruses and bacteria could easily pass through a direct filtration plant, despite significant microorganism removal. Results also identified those operational variables which enhanced the removal of bacteria in the direct filtration process (Hefta 1990). Therefore, disinfection processes would still have to be employed to approach the goal of zero biota transfer. In most cases, by law, surface waters used for drinking water require some pretreatment, like direct filtration, before maximum efficiency can be achieved in a disinfection process.

A pilot scale model of a filtration facility was constructed and used in three different phases of testing. During the first phase, the pilot plant was operated at the University of North Dakota Hydraulics Laboratory. Tap water amended with chemicals and clay was used to simulate Garrison Diversion water. Chemical doses and the suspended solids load were varied to evaluate the general performance of the pilot plant. In the second phase, some modifications were made to the pilot plant, and it was moved to Lake Sakakawea to evaluate direct filtration of actual Garrison Diversion water. During phase three, the pilot was returned to the University of North Dakota where more tests were performed to evaluate different chemicals to aid in solids and bacteria removal.

The pilot scale plant used at Lake Sakakawea is shown in Figure 9.1. Raw water is pumped into a rapid mix basin where chemicals like alum (coagulant) or synthetic polymers (coagulant aids) can be added. The purpose of the rapid mix basin is to blend chemicals thoroughly with the incoming water. The detention time in the rapid mix basin was approximately 5 minutes (Hefta 1990). Coagulants chemically act on very fine particles and some dissolved compounds to form particles that are large enough to settle by gravity. This process is known as flocculation. The larger particles are called flocs. Polymers are large branch-like molecules with chemically reactive sites along the branch. The floc attach to the polymer to make even larger particles that are easier to remove during filtration. Following flocculation, water is pumped to filters where floc is removed by filtration mechanisms. The filter media consisted of 50 centimeters (20 inches) of coal over 25 centimeters (10 inches) of sand. Below the sand was 25 centimeters (10 inches) of graded sand and gravel that served as the underdrain system.

The influent and effluent water were tested for turbidity (an indirect measure of suspended solids), total heterotrophic bacteria, and, in some instances, the population of *Yersinia ruckeri*. Other data recorded included head loss measurements. The head loss measurement across the filter media is necessary to predict the total time a filter could operate before backwashing, and therefore, the total amount of filtered water that would have to be used to backwash the filters, compared to the total amount of water treated. Other observations included differences in performance due to different doses of chemicals and polymers, and different detention times in the flocculation basin.

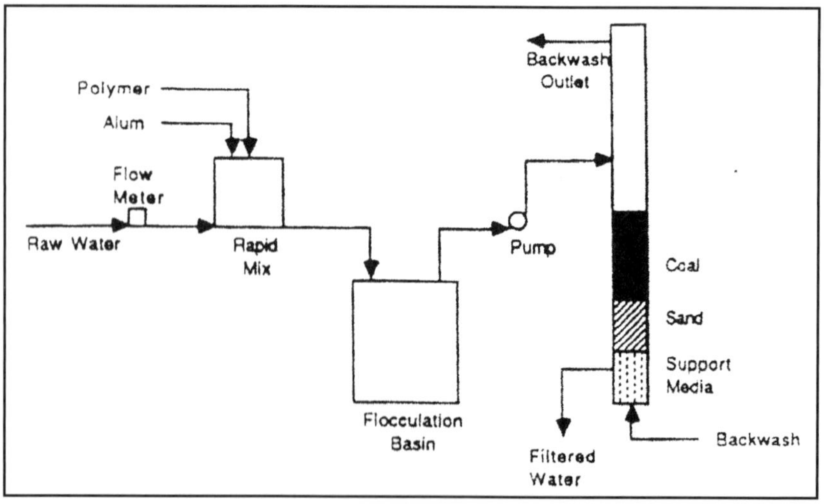

Figure 9.1. Pilot scale plant used at Lake Sakakawea.

Results of the research from 1990-1992 study indicated that Lake Sakakawea water can be treated to drinking water turbidity standards by direct filtration using alum as a coagulant and with or without using flocculation. (Moretti et al. 1993). The amount of filterable material in Lake Sakakawea water was low enough that the water could be filtered at high rates for over 24 hours before backwashing was required. This included testing of lake water in late summer when algae that tend to clog filters should be at maximum levels. The filtration rate could be 0.306 cubic meters per minute per square meter (7.5 gallons per minute per square foot) or higher; with the long filter runs, observed, direct filtration could be used to treat Garrison Diversion water (Hefta 1990). During several sets of filter runs the bacterial removal was assessed (50-90 percent). In general, without polymer addition, bacterial removal decreased with increasing flow rates and decreasing alum dosage. More importantly, without flocculation, it was difficult to achieve high bacterial removals at high flow rates or low alum doses.

The results of this research were significant for three reasons. First, it was apparent that although direct filtration could reduce the turbidity to levels acceptable for drinking water, it could not prevent the transfer of biota like viruses and bacteria. Second, it showed that consumers within the Missouri River watershed could use direct filtration to treat their drinking water supply. Finally, the data generated were an excellent starting point for any additional research or application of direct filtration of Garrison Diversion water.

As a part of continued research at the University of North Dakota, direct filtration was further evaluated in addition to investigating disinfection options (Moretti and Kopchynski 1992; Moretti et al. 1993). Direct filtration testing consisted of several filter runs in the laboratory, and at three different times during the year at Lake Sakakawea. The pilot plant model used during this study had the same components as the 1988-1990 study. However, this work expanded the previous study to include an evaluation of three different media configurations, several types of polymers, and the effect of the hydraulic flow rate. Several runs also included bacterial removal determinations. Based on the direct filtration pilot plant results, a conceptual design and a preliminary cost estimate was presented.

The three media configurations included a dual media bed (sand below coal); a multimedia bed with sand, coal and ilmenite; and a deep uniform coal bed. The underdrain system again was graded sand and gravel. Laboratory tests conducted with synthetic Garrison Diversion water (similar to the one used in the previous study) indicated that the dual media configuration was more efficient than the other two configurations. The multimedia bed exhibited rapid clogging, which led to short run times before backwashing was required. Detailed examination of the head losses in the different layers indicated that the increased head loss was caused by the ilmenite layer. This was attributed to the very small, closely packed grains that compose ilmenite. Because high flow rates and long run times before clogging are desirable for direct filtration applications such as the one proposed for the Garrison Diversion Unit, the multimedia configuration was not tested at Lake Sakakawea.

The deep bed configuration consisted of 152 centimeters (5 feet) of coal. Head losses were very low for this configuration, even at hydraulic loadings of 0.55 cubic meters per minute per square meter (13.5 gallons per minute per square foot) which is nearly twice the flow rate used in the previous study. Unfortunately, filter runs were still very short because solids quickly penetrated the bed and appeared in effluent from the filter. Once suspended solids increased enough to cause turbidity to rise above the acceptable drinking water level, filters had to be stopped and backwashed.

The dual media configuration resulted in the best performance. The dual media bed consisted of 41-46 centimeters (16-18 inches) of coal above 15-20 centimeters (6-8 inches of sand). The combination of larger coal particles above the smaller sand resulted in long run times without excessive head loss or suspended solids breakthrough. The conceptual design calls for 15 centimeters (6 inches) of sand below 46 centimeters (18 inches) of coal.

There are several chemicals that may be used as coagulants and coagulant aids. Based on years of experience, alum is preferred as a coagulant. Different polymers work differently as coagulant aids in different waters, depending on compounds present in the water. Five different polymers and alum were tested to determine what type of polymer, and what doses of polymer and alum achieved the best filtration performance with the dual media configuration. Polymers were selected so low molecular weight, medium molecular weight, high molecular weight, and ultra high molecular weight types were all tested. Molecular weight is important because it can determine the size of the flocculated particles formed. Large molecular weight polymers are generally longer, more branched, molecules with more surface reaction sites for coagulated particles to attach. Therefore, large molecular weight polymers should result in larger floc particles. Different polymers also have different chemical charges. The polymers tested included cationic polymers, which have positive charges and attract negatively charged particles, and nonionic polymers which contain both positive and negatively charged sites.

Results indicated, in general, the low molecular weight polymers outperformed the high molecular weight polymers on Garrison Diversion water. This was due to high molecular weight polymers tending to form larger flocs which clogged the upper portions of the media, restricting flow before the entire bed depth had been utilized. This resulted in rapid increases in head loss and short filter runs before backwashing. Other advantages of low molecular weight polymers were easy blending into raw water and very low doses required to produce very high quality filter effluent. Both of these aspects are extremely important. To obtain the maximum effect, chemicals added must be completely blended into the water. Otherwise more chemical must be added to achieve the same result. Poor blending results in greater chemical consumption and cost. For example, during testing, some filter runs were performed without a rapid mix step. In these instances performance suffered. Furthermore, the chemical that produces the best results with the lowest dose is usually the most cost effective based on raw chemical cost and storage requirements. Low molecular weight, cationic polymer (at a dose of 0.03 milligrams per liter) worked well. It was also suggested that alum need not be added.

Several important points need to be considered concerning the coagulation and flocculation of waters. First, different doses and different types of chemicals are effective in different types of water, depending on the nature of the compounds in the water and, to a great degree, on the pH, or acidity, of the water. The optimum dose should be determined based on what dose results in the best quality effluent and longest filter runs. Chemical addition beyond the optimum may result in poorer performance. Chemicals represent one of the most significant operational costs of water treatment plants. This is not significant in direct filtration plants.

One problem in treating surface waters is that water quality changes from season to season and year to year. Differences in water chemistry and quality may demand adjustments in chemical doses, and the type of chemical used. A major problem for direct filtration systems is the presence of algae. Heavy algae growths are usually a seasonal phenomenon and may result in very short filter runs due to excessive clogging. The effect of algal blooms may only be partially compensated for by adjusting the chemical feed. Algae currently prevalent in Lake Sakakawea water are not the most problematic species for direct filtration. Both the 1988-1990 and 1990-1992 investigations included filter runs on late summer samples of lake water and showed that the presence of algae in those summers did not prevent effective filtration, although filter runs were shorter. It was also pointed out that Lake Audubon water, and, hence, McClusky Canal water have historically had significantly higher algal counts than Lake Sakakawea (Hefta 1990). This could lead to problems applying the available data to a direct filtration plant on the McClusky Canal. Higher levels of dissolved organics in Lake Audubon may also affect the results of direct filtration tests.

Another important aspect of seasonal variations is temperature. A direct filtration plant for the Garrison Diversion Unit may encounter water below 4°C. Chemical reactions like flocculation proceed at much slower rates at lower temperatures. If flocculation is desired for filtering Garrison Diversion water, flocculation time may have to be increased. This could be done by lowering flow rates through the plant, thereby increasing the residence time of water in the flocculation basin,

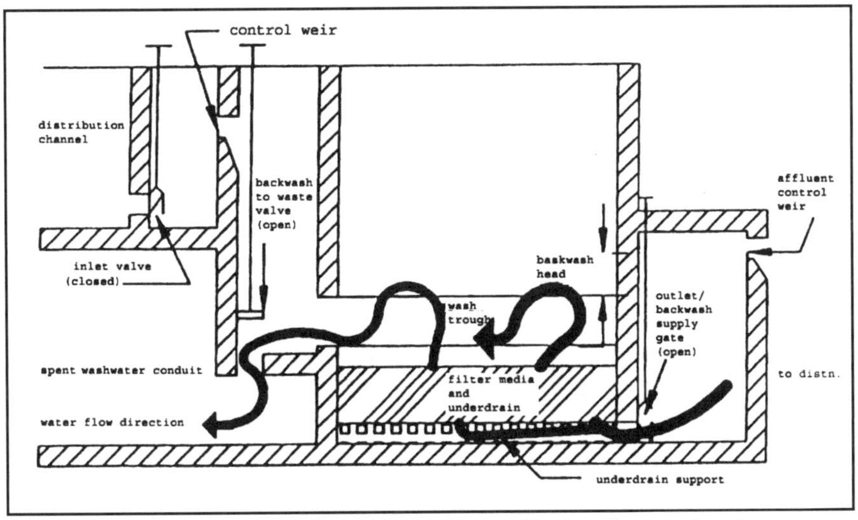

Figure 9.2. The proposed filter plant with the mode of backwash operation shown.

but reducing facility capacity. It was also noted that flocculation did occur within the filter media when no flocculation basin was included in some pilot plant filter runs (Hefta 1990). Whether this is true for cold temperatures was not clear. It may also be argued that it is possible to have flocculation basins sized for warm conditions and rely on flocculation within the filter media in cold weather. Unfortunately, direct filtration plants often experience problems at low temperatures because the flocculation continues to grow slowly within the filter media (Hefta 1990). This can lead to excessive or unpredictable clogging; generally, an extended flocculation time is necessary for effective filtration at low temperatures.

The results from bacterial population counts indicate summer removals averaged 90 percent, while in colder weather, some runs appeared to have no removal. In some cases, water samples were stored for 3-5 days before plating and counting. In addition to the potential regrowth problem, it is difficult to assure sterile conditions and evaluate bacterial removal for separate filter runs. In between runs, it is likely that bacteria can grow on the media surface, or even inside the coal media. Therefore, it would be difficult to determine the total bacterial removal for individual runs because some residual bacteria from a previous run may breakthrough on the run being tested. Despite these difficulties, it was concluded that, under optimum conditions, direct filtration could probably not consistently remove more than 90 percent of the bacteria from raw water (Moretti et al. 1993).

Based on the results from both sets of pilot studies, the conceptual design of a direct filtration plant (Moretti et al. 1993) calls for a low molecular weight, cationic polymer dose of 0.03 milligrams per liter added to raw water and forced through a high velocity nozzle to thoroughly mix the polymer with the raw water. From there, the water is fed directly to the filters, probably through an open trough or multiple troughs. The design does not call for flocculation basins. It was assumed that adequate flocculation will occur during travel time between the nozzles and the filter media. In this case, it was stated that low degrees of flocculation would be desirable because small particles would filter better than the large particles.

The suggested filter media configuration was 46 centimeters (18 inches) of coal over 15 centimeters (6 inches) of sand. Filters were sized assuming a hydraulic loading of 0.306 cubic meters per minute per square meter (7.5 gallons per minute per square foot) and a total daily flow of 1,892,500 cubic meters per day (500 million gallons per day). With this hydraulic loading and total flow, four sets of ten filters would be required. Each filter bed would be 35 feet long, 35 feet wide, and 22.5 feet deep. Provisions for a small chlorination facility were also included in filter design. This would help control any growth of algae or bacteria within the filter plant. The flow scheme is shown in Figure 9.2.

Backwashing filter media usually requires pumps to move water from the clear wells through underdrain system and into the media. The backwashing pumps make up a significant initial and annual cost. The conceptual design calls for a gravity backwash system to eliminate the need for backwash pumps. When the inflow of water to the filters is stopped in gravity backwash systems, filtered water flows by gravity back through the underdrain. The gravity backwashing mode is also depicted in Figure 9.2. The design calls for spent backwash water to be stored in lagoons.

There are two potential drawbacks to the proposed design. Frequently after backwashing, filtered water has a lower quality than after the filter has run for several minutes. During backwashing all deposited solids in the filter are not backwashed. Some remain in the filter and will be flushed out with the filtered water when filtering is resumed. In addition, as the filter slowly fills with retained solids, the straining, interception, sedimentation, and flocculation processes may be enhanced because there is more surface area within the filter media. This is called filter ripening (Hefta 1990). It may be desirable to add piping and valves to allow the first few minutes of filtered water to pass through the filter, and either be discharged to the backwash water lagoon or be returned to the head of the plant.

Another drawback noted in the design is high head loss, or pressure drop, across the filters. Water moves through the filter media by gravity. To filter the large flow rates, the elevation of the water on the influent side of the filters needs to be much higher than at the outlet channel. To accommodate the high head loss, it was suggested the filtration plant be located below a dam, or drop structure, to avoid pumping costs.

Direct filtration of Garrison Diversion water, especially if flocculation with a synthetic polymer was used, would help remove bacteria, viruses, and protozoans, as well as be an effective barrier against fish, fish eggs, and fish larvae. However, like the fish screen, a direct filtration facility cannot reliably remove all viruses and bacteria. Investigators of direct filtration of Garrison Diversion water have noted direct filtration would have to be coupled with a disinfection process (Turner and Hefta 1989; Hefta 1990; Moretti et al. 1993). For disinfection to be effective, turbidity of the water being disinfected must be reduced to very low levels. Even very small particles shield bacteria and viruses from disinfectants. The phenomenon is not easily explained. However, what is clear is that bacteria tend to grow well on surfaces; and when solids are present, bacteria are extremely difficult to remove by disinfection, even at very high disinfectant doses. If disinfection is required to prevent biota transfer, direct filtration will likely be necessary before disinfection to reduce the turbidity to drinking water levels. Furthermore, direct filtration, if coupled with polymer addition, successfully reduces the level of viruses and bacteria in treated water. Therefore, coupling direct filtration with polymer addition and disinfection will make disinfection more effective and provide two barriers to biota transfer.

Disinfection

Chlorine and chlorine-based compounds have been the most popular disinfectants. Concerns over environmental and public health effects of chlorine are making ozone and UV (ultraviolet light) more popular in new and retrofitted water and wastewater plants. Chlorine, ozone, and UV disinfection have all been considered for use in the Garrison Diversion Project. Although all three disinfectant options have proven feasible for disinfecting hatchery waters to control *Yersinia ruckeri* and IHNV, their usefulness in biota removal in interbasin transfer of water is yet to be proved (Wedemeyer and Nelson 1977; Bullock and Stuckey 1977; Wedemeyer et al. 1978; Sako and Sormachi 1985).

In the eastern part of the Garrison Diversion Unit, treated water would likely be returned to a reservoir or canal system for irrigation, rural and municipal water supply, and recreation in the Sheyenne and Red River watersheds. Another part of the Garrison Diversion Unit involves transferring water to the Hudson Bay watershed via the Souris River watershed. The proposed Northwest Area Water Supply Project area includes a pipeline to transfer water from Lake Sakakawea directly to Minot's (North Dakota) water treatment plant. Current plans call for ozone or chlorine disinfection in the pipeline without prior filtration. Some concerns obviously have been raised over the potential for high turbidities, resulting in poor disinfection within the pipeline. In this instance, however, water would be piped to a complete water treatment plant that already includes filtration and disinfection processes. Therefore, this scheme also includes multiple barriers to biota transfer.

Ultraviolet light disinfection

Ultraviolet light is becoming more popular in water and wastewater plants in North America. Some reasons for UV's popularity include its excellent action against viruses, high reliability, and a lack of a chemical residual (Water Pollution Control

Federation 1984). Unlike chlorine and ozone, UV light is a physical process that does not rely on adding a chemical to the water. UV light is capable of altering chemicals within a water sample; but at the low power used to disinfect water and wastewaters, it is generally assumed that UV light will not create significant amounts of harmful chemical byproducts (Moretti et al. 1993).

UV light is generated by passing an electric current through mercury vapor in a low pressure bulb that is surrounded by a quartz sleeve. The electric current causes mercury vapor to release UV radiation. The quartz sleeve serves to protect the lamps and insulates the lamps from heat loss (Water Pollution Control Federation 1986). Banks of sleeved lamps are typically submerged in a channel where water flows past the lamps. The number of lamps, lamp intensity, and hydraulic residence time in the channel are the critical design parameters.

UV light owes its disinfecting power to the relationship between the specific wavelength of UV light emitted by the bulbs and how well the microbes' DNA absorb UV light. DNA has its greatest absorbance at 256 nanometers (0.000000256 meters or 0.000010079 inches) which is in the ultraviolet region of the electromagnetic spectrum (Water Pollution Control Federation 1984). Therefore, DNA absorbs ultraviolet light. When DNA absorbs UV light, proteins within the DNA are damaged, making growth and reproduction impossible.

The most important restriction with UV light for disinfection is the quality of the water (Metcalf and Eddy 1979). When light passes into water, suspended solids tend to scatter the light, and dissolved organics will absorb some of the light. This decreases the amount of UV light that can reach target organisms. Coagulation, flocculation, and direct filtration processes will lower both suspended and dissolved solids, making UV disinfection more effective. Lake Sakakawea water is typically very low in suspended solids which makes it an excellent candidate for UV disinfection. As a part of the investigations at the University of North Dakota, the transmittance of filtered and unfiltered Lake Sakakawea water samples was determined to ensure adequate light penetration (Moretti et al. 1993; Moretti et al. 1996). This measurement was made by passing UV light (254 nanometers) through 1 centimeter of lake water. The transmittance of both the filtered and unfiltered samples was 81 percent, well above the minimum suggested for UV disinfection, confirming that even unfiltered Lake Sakakawea water could be treated with UV light. Water in Lake Audubon and the McClusky Canal could have higher contents of both suspended and dissolved solids due to higher populations of algae and other plankton, which could reduce UV light transmittance.

Laboratory tests were conducted using a standard UV lamp to disinfect filtered and unfiltered Lake Sakakawea water samples. Water samples were exposed from 15 seconds to 2 minutes. The reduction in natural lake bacteria and *Yersinia ruckeri* was determined to evaluate the feasibilty of a UV disinfection system. *Yersinia ruckeri* (332,000 microorganisms per milliliter) was completely inactivated by the single lamp with an exposure time of only 30 seconds in an unfiltered water sample. Other tests showed that UV disinfection was also able to significantly reduce background bacteria normally present in Garrison Diversion water, even at short exposure times (Moretti et al. 1993; Moretti et al. 1996).

The conceptual design for a full-scale facility would need 20,160 UV lamps (147 cm) to be submerged vertically in 18 channels. Each channel would be 1.55 meters (61 inches) deep, 1.27 meters (50 inches) wide, and 11.3 meters (37 feet) long. Each channel would have a capacity of 113,550 cubic meters per day (30 million gallons per day), and water would travel at 0.7 meters per second (28 inches per second) through the channel.

In this study, water quality was known, water was stagnant, and exposure time was controlled. However, the lamp intensity was not known. Therefore, the results only confirmed that *Yersinia ruckeri* and other microorganisms naturally present in lake water could be disinfected by UV light. The design was based on literature values of lamp power, travel time (exposure time), and the fact that Lake Sakakawea water was suitable for UV disinfection.

There were several concerns expressed over the design. Perhaps most importantly, UV disinfection systems are usually used only in small plants, which are much smaller than the 1,892,500 cubic meters per day (500 million gallons per day) proposed for the Garrison Diverison Unit. It also is not clear how often the lamps will have to be cleaned and/or replaced. In some facilities, cleaning is only necessary when films of microorganisms begin attaching to the exterior of the lamps and, in some cases, scaly deposits form on the lamps, diminishing the intensity of the delivered light. The lamps will periodically have to be replaced. This adds to maintenance required for the disinfection. It is not clear whether winter operations will lead to ice, which could severely damage the bulbs. One of the most important conclusions concerning the feasibility of UV light for the Garrison Diversion Unit application was that significant pilot plant testing would have to be performed to clarify the uncertainties (Moretti et al. 1993).

Some microorganisms can repair damaged DNA. The damage to DNA caused by UV light may also be chemically repaired if the extent of injury is not extreme. When only injured, organisms may repair damaged DNA in sunlight, fluorescent, or incandescent light in a process called photoreactivation (Water Pollution Control Federation 1986). The damage caused by UV light cannot be fully repaired; typically only a few injured organisms will actually repair the damaged proteins (Water Pollution Control Federation 1984). Furthermore, only a few bacterial strains have the appropriate enzyme system to carry out the repair. Photo reactivation poses a problem because injured cells would not grow during microbiological tests used to examine the water immediately after treatment, but would grow if the water was tested after some period of time. In some cases, at wastewater treatment plants, discharge permits must account for a certain number of inactivated cells being capable to sufficiently repair injury to survive the UV disinfection process (Water Pollution Control Federation 1986). This poses a potentially severe limitation for Garrison Diversion water if *Yersinia ruckeri* is capable of photoreactivation .

Chlorination disinfection

Chlorine compounds have traditionally been the most popular disinfectants for water and wastewater plants. Chlorine owes its disinfecting action to its strong chemical oxidizing power. Chlorine reacts with proteins, destroying the biochemical machinery required for microorganisms to proliferate (Viessman and Hammer 1985). As a disinfectant, chlorine has proven effective.

Chlorine can be introduced into water from several compounds, such as chlorine gas, hypochlorites, and chlorine dioxide. The conceptual design of a chlorination facility for the Garrison Diversion Unit calls for liquefied chlorine gas to be supplied in railroad tanker cars (Moretti et al. 1996). Using other forms of chlorine eliminates much of the concerns over worker safety, but poses other problems.

When sufficient quantities of gaseous chlorine or hypochlorites are added to water, they form hypochlorous acid (HOCl) in equilibrium with its dissociation product hypochlorite (OCl^-). The equilibrium is controlled by chlorine, concentration, pH, and temperature. A helpful feature of the feed chlorine compounds are their high solubilities. Therefore, when hypochlorites or chlorine gas are added to water, they will readily dissolve.

Hypochlorous acid and the hypochlorite ion are called free chlorine residual. These are both relatively strong disinfectants, with hypochlorous acid being many times stronger. They are called free chlorine residual. Because of its very strong oxidizing power, chlorine reacts very quickly with various compounds in water which can result in chlorinated byproducts with much lower disinfecting power. The chlorinated byproducts are called combined chlorine residual. Compounds in water that will react with chlorine, preventing the formation of a lasting free chlorine residual, include ammonia, organic nitrogen, reduced metal species like ferrous iron, manganese, other halogens like bromide, and various organics. When reactions between chlorine and these compounds produce chlorinated compounds, especially chloramines (a combination of ammonia and chlorine) and chlorinated organics, the combined residual developed has only a limited disinfecting power.

All the compounds that react with chlorine before a combined or free chlorine residual is developed contribute to what is known as chlorine demand. Chlorine demand is the difference between the amount of chlorine added and the sum of the combined and free chlorine residual remaining in the water. When the chlorine demand is high, more chlorine must be added to establish a free or combined residual that can disinfect the water being treated. The chlorine demand of Lake Sakakawea water samples is low compared to many other surface waters (Moretti et al. 1993; Houston Engineering et al. 1996). Low demands translate to savings in annual chemical costs. Water samples from Lake Audubon exhibited a slightly higher demand (Houston Engineering et al. 1996), likely due to greater concentrations of dissolved organic matter.

One of the most important disadvantages of using chlorine relates to the safety of persons handling and storing chlorine, and residents that may live near a chlorination facility. Chlorine gas is extremely corrosive and, if inhaled, forms hydrochloric acid in human lungs which could be fatal. There are significant safety measures required for handling chlorine gas. To prevent harm to residents and water plant personnel, chlorine gas must be stored in a building isolated from the rest of the water treatment plant. This building must be equipped with monitoring equipment, have restricted access, and may require air pollution control equipment to clean up any leaks. These safety measures add to the costs of chlorinated facilities.

Chlorinated organic compounds have been implicated in cancer in humans, and chloramines have been shown to be toxic to aquatic life. Because disinfected Garrison Diversion water, with the exception of the Northwest Area Water Supply Project, will likely be discharged to canals or reservoirs, residual chloramines and chlorinated organics should be a major concern if chlorine disinfection is employed.

Residual free chlorine and combined residual can be removed from disinfected water by adding sulfur dioxide gas or filtrating through activated carbon. In most cases, wastewater treatment plants dechlorinate by adding sulfur dioxide gas. The design for the chlorination unit calls for sulfur dioxide dechlorination, which will represent a significant percentage of the cost of a chlorination facility (Moretti et al. 1993). Adding sulfur dioxide gas typically does not create any detrimental byproducts; but if the dose is too large, it can deplete the oxygen concentrations in the treated water. This would pose a problem to aquatic life in the canals and reservoirs receiving dechlorinated Garrison Diversion water if prolonged periods of overdosing or any accidents or malfunctions occurred in the dechlorination process. Therefore, a reliable system for monitoring chlorine residual and adjusting the required sulfur dioxide dose, as well as safeguards against accidents, is critical.

Other chlorinated compounds that pose health concerns are trihalomethanes and haloacetic acids. Trihalomethanes are formed when organic compounds resulting from the decay of vegetation, like humic acids, come into contact with chlorine. The level of these disinfection byproducts is an obvious concern for the Minot, North Dakota, portion of the Northwest Area Water Supply Project because the water will be directly used for water supply. Trihalomethanes may pose problems for the Garrison Diversion Unit in eastern North Dakota because they are difficult to remove from waters once they are formed. Although they can be removed by activated carbon filtration and aeration, the best method to control trihalomethanes in drinking water is to prevent them from forming.

To prevent trihalomethane formation, organic compounds that are the precursors to trihalomethane formation need to be removed from the water before chlorine is applied. The water treatment practices of coagulation-flocculation and filtration through charcoal media will serve to reduce the level of these organic precursors, but to what extent this is true for Garrison Diversion water is not clear. The levels of trihalomethane and haloacetic acid formation of unfiltered Lake Sakakawea and Lake Audubon waters were evaluated during chlorination tests conducted for the Northwest Area Water Supply Project (Houston Engineering 1996). The levels did not exceed drinking water standards, but it was noted that additional contact with chlorine at the Minot water treatment plant could further increase the levels of these harmful compounds.

Besides the chlorine demand of the influent water, and health concerns related to chlorine chemicals and byproducts of chlorine disinfection, several other factors will influence the ability of chlorine to prevent biota transfer. Two obvious variables are pH and temperature, which influence chemical equilibrium and chemical reaction rates. Cold temperatures slow down chemical reactions between chlorine compounds and microorganism enzymes. With other factors being equal, this increases the time necessary to disinfect the same amount of microorganisms in warmer waters. The pH of the water affects basically all chemical reactions in water. In the case of chlorination, a lower pH results in a higher disinfecting power. This is due to the equilibrium between hypochlorous acid and the hypochlorite ion. At a low pH, more of the chlorine is present as hypochlorous acid which is the stronger disinfectant. This may be important if Garrison Diversion water is ever softened before being disinfected and transferred across the watershed boundary. After softening, water has a very high pH. Therefore it will be beneficial to reduce the pH before disinfection to increase the disinfecting power of chlorine. Temperature also affects the equilibrium between hypochlorous acid and the hypochlorite ion, but will likely be overshadowed by other factors.

Other factors influencing disinfection efficiency relate both to chlorination and ozonation. These factors include the type of microbes being inactivated, the condition of the microbes, the hydraulic condition of the reactor where the microorganisms are exposed to the disinfectant, the design of the equipment used to feed the disinfectant into the water, the contact time, and the concentration of the disinfectant used.

There are a wide variety of conditions that make microorganisms more or less susceptible to chemical disinfectants. This may range from the temperature and nutritional growth conditions prior to disinfection to sublethal injury or acquired resistance due to prior exposure to disinfectants. Different microorganisms can respond differently to these factors depending on the type and age of the microorganisms and numerous other conditions (Kopchynski 1991).

Another important factor has a more distinct response. When microorganisms are present in clumps of flocculent or grainy biomasses, they are more difficult to disinfect (Kopchynski 1991). This can be related to the shielding effect caused by turbidity. When attached to other bacteria or inert solids, microbes tend to be protected from disinfectants.

The resistance of different microorganisms to disinfectants varies based on the type of microbe, the type of disinfectant, temperature, and pH. In general, actively growing bacteria are not very resistant, while bacterial spores, viruses, and protozoan cysts like those of *Giardia lamblia,* are much more resistant. Disinfectants have different efficiencies against different types of microorganisms. For example, ozone is a much stronger disinfectant than chlorine, but the difference between the bacteridal and viricidal doses tends to be much smaller for ozone than chlorine (Moretti et al. 1996). This means that ozone is considered more effective against viruses than chlorine; when using ozone, less additional disinfectant is required to inactivate the viruses over bacterial inactivation than would be required with chlorine.

Since the disinfection depends on C (milligrams per liter), the residual concentration of the disinfectant, and t (minutes), the exposure time, the product Ct is used as a measurement of disinfection capability. The EPA has published tables of Ct values required to disinfect surface waters used for drinking water against given percentages of *Giardia* cysts and enteric viruses (U.S. Environmental Protection Agency 1990). Ct values are a function of the type of disinfectant, temperature, and pH. For ozone, there is relatively little pH effect on disinfection, so pH is not included in their Ct values. Inspection of the tables reveals that the strongest disinfectant, ozone, requires lower Ct values than the weaker disinfectants, free chlorine and chloramines, to inactivate the same percentage of microorganisms.

Two approaches can be taken to determine the Ct values in laboratory experiments. The inactivation tests performed with chlorine and ozone at the University of North Dakota used a fixed residual concentration over a given period of time (Hefta 1990; Cruise 1993; Moretti et al. 1993; Moretti et al. 1996). Chlorine tends to establish a relatively constant residual over at least several hours because the chlorine demand is generally satisfied very quickly and chlorine is present as dissolved species in the water. However, ozone is a gas that after dissolving in water quickly degrades to oxygen, leaving no lasting residual. Therefore, the ozone concentration declines rapidly over a few minutes unless ozone is continuously supplied to stagnant water. In the case of the studies conducted at the University of North Dakota, a constant ozone residual was maintained so the Ct value is simply the product of disinfectant concentration and contact time (Cruise 1993; Moretti et al. 1993; Moretti et al. 1996). Different testing procedures were used in North Dakota State University (Richard et al. 1996) and the Northwest Area Water Supply Project (Houston Engineering et al. 1996) to determine Ct value from the experimental data. A typical ozone residual decay over time is shown in Figure 9.3. The area under the curve in this figure is the Ct value for the experiment. The experimental data points were fitted to an exponential decay model and a linear decay model. The difference between the methods used to determine ozone and chlorine Ct values is clearly established by examining behavior of residuals in actual contact basins used to disinfect water. In chlorine contact basins, the water flows through a baffled chamber like the one shown in Figure 9.4. Because chlorine establishes a relatively stable residual as shown in Figure 9.5 for samples from Lake Audubon and Lake Sakakawea, the product of the contact time and the residual provides an accurate estimate of Ct value. In ozone contact basins like the one depicted in Figure 9.6, ozone is usually applied in the first compartment of the basin. As the ozone decays to oxygen in the following compartments, the residual declines (Figure 9.3). In most waters, the ozone residual decays exponentially (Richard et al. 1992). The Ct values in water from Lake Sakakawea and Lake Audubon were determined assuming an exponential ozone decay in the NAWS study. During the ozone inactivation tests at North Dakota State University, an exponential model provided the best fit for the experimental inactivation of *Y ruckeri* in lake water; but for the inactivation in phosphate buffered saline, the linear model provided a superior fit to the experimental data (Richard et al. 1996).

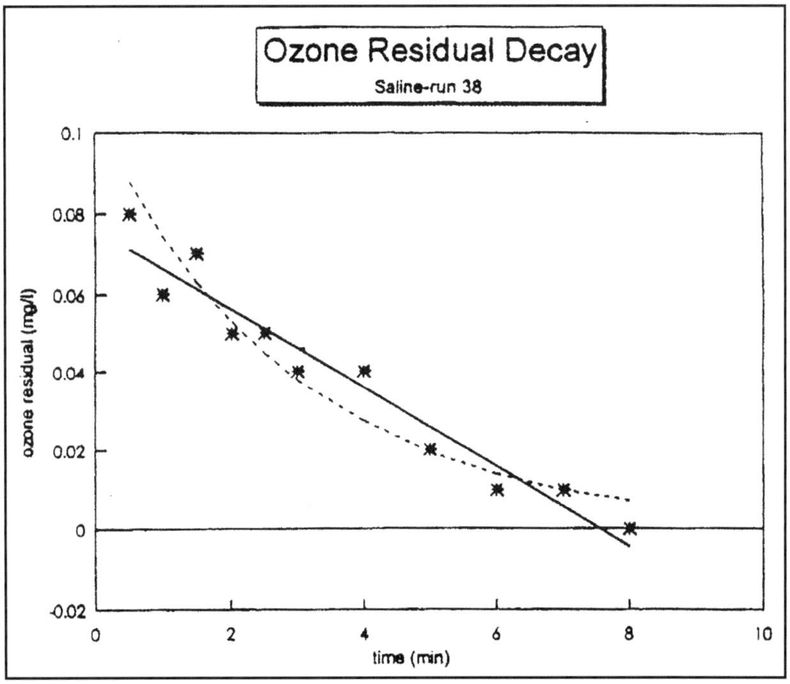

figure 9.3. Residual ozone decay over time.

Chlorine demand and inactivation tests with *Y. ruckeri* and lake bacteria in filtered and unfiltered water from Lake Sakakawea were performed during the research conducted at the University of North Dakota (Hefta 1990; Moretti et al. 1993). NAWS research reported free chlorine and chloramine inactivation of *Giardia muris* cysts and the MS2 virus which infects *Eschirichia coli* bacterium (Houston Engineering et al. 1996). In addition to the inactivation tests, the formation of haloacetic acids and trihalomethanes were also evaluated.

The initial direct filtration study concluded that some sort of disinfection was required to inactivate the microbe remaining after direct filtration (Hefta 1990). Significant details were not provided, but it was reported that a dose of 0.1 mg/l of chlorine with a contact time of 10 minutes could inactivate bacteria remaining in filtered water. This information showed that it was feasible to use chlorine and, hence, other disinfectants to eliminate microbes remaining in filtered Lake Sakakawea water. Important information that was not determined included chlorine demand and measurements of free and combined chlorine residual.

Chlorine demand tests were conducted during the next phase of research at the University of North Dakota (Moretti et al. 1993). Filtered and unfiltered Lake Sakakawea samples from July, October, and February were tested. Sodium hypochlorite was added to samples, and the total chlorine residual was monitored after 20 minutes. Chlorine demand averaged 0.8 milligrams per liter and never

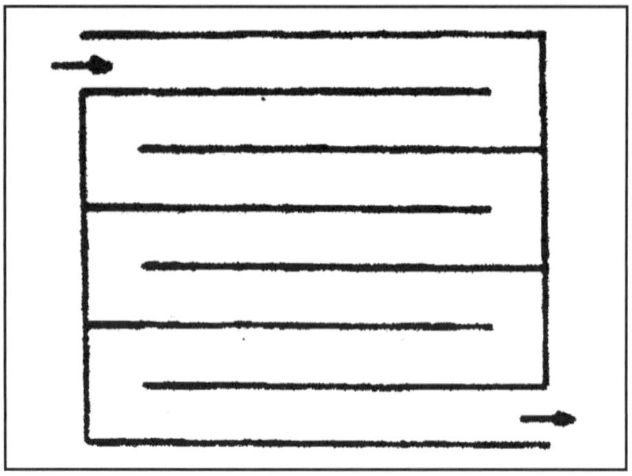

Figure 9.4. Longitudinal baffles in the chlorine contact basin.

exceeded 1.0 milligrams per liter. There was no direct relationship between sample turbidity and chlorine demand. Some unfiltered samples, with turbidities as high as 19 NTUs (nephelometric turbidity units), had lower demands than some filtered samples with turbidities as low as 0.32 NTUs. Although turbidity may have a detrimental impact on inactivation effectiveness, it appears that high turbidities did not significantly affect the chlorine demand of the water samples. Results also suggest that most of the changes in chlorine demand are related to dissolved organics or the type of dissolved solids present.

The demand tests conducted for the Northwest Area Water Supply Project supported the results from the University of North Dakota. Five samples from Lake Sakakawea and Lake Audubon again revealed that the average chlorine demand after 10 minutes was 0.7 milligrams per liter and 1.0 milligrams per liter, respectively. Again, the higher demand for Lake Audubon again was likely due to increased levels of dissolved organics.

Chlorine inactivation tests at the University of North Dakota were performed on filtered and unfiltered samples for total bacteria and on a filtered sample for *Y. ruckeri*. Initial total bacterial counts ranged from 10,000 to 634,000 microorganisms per milliliter, and the tests were conducted at 14°C. Tests on filtered water with turbidities less than 0.9 NTUs showed that a free residual of 1.25 inactivated 99.8 percent of the lake bacteria within 20 minutes. In the single unfiltered sample tested, a free residual of 1.1 mg/l with a contact time of 20 minutes inactivated only 89.6 percent of total bacteria. This illustrates the effect of turbidity on the effectiveness of disinfection. The turbidity of this sample was 6.7 NTUs, which supports the premise that direct filtration prior to disinfection is desirable, especially if small doses of chlorine are to be used for disinfection.

Inactivation tests performed on a sample of filtered water spiked with the *Y. ruckeri* bacterium showed that *Y. ruckeri* was more sensitive to chlorine than other bacteria in the water. Tests were conducted at temperatures of 2 and 20°C, with contact times ranging from 1 to 20 minutes. The turbidity of the filtered sample was 0.5 NTU, and the initial *Y. ruckeri* concentrations were about 50,000 microorganisms per milliliter. Results revealed that a free chlorine residual of 0.5 mg/l was capable of essentially eliminating *Y. ruckeri* from filtered water within 10 minutes at either 2 or 20°C.

The Northwest Area Water Supply Project targeted both the MS2 bacterial virus and *Giardia muris* for inactivation in unfiltered water from Lake Sakakawea and Lake Audubon. Inactivation tests were conducted at 4°C. Initial concentrations of MS2 virus and *Giardia muris* were approximately 1,000,000 and 10,000 microbes per milliliter respectively. In these experiments, a free chlorine residual was developed for a period of time, followed by the addition of ammonia to turn the free residual into a combined residual consisting of chloramines. The target chlorine residual was 4 to 4.5 milligrams per liter.

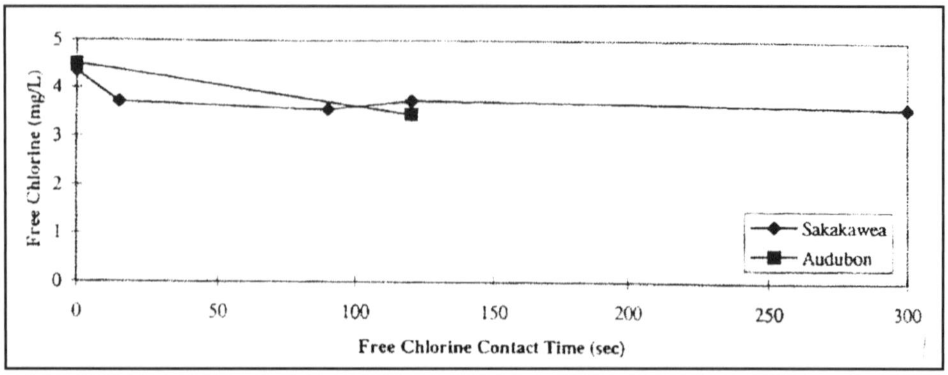

Figure 9.5. Residual chlorine over time.

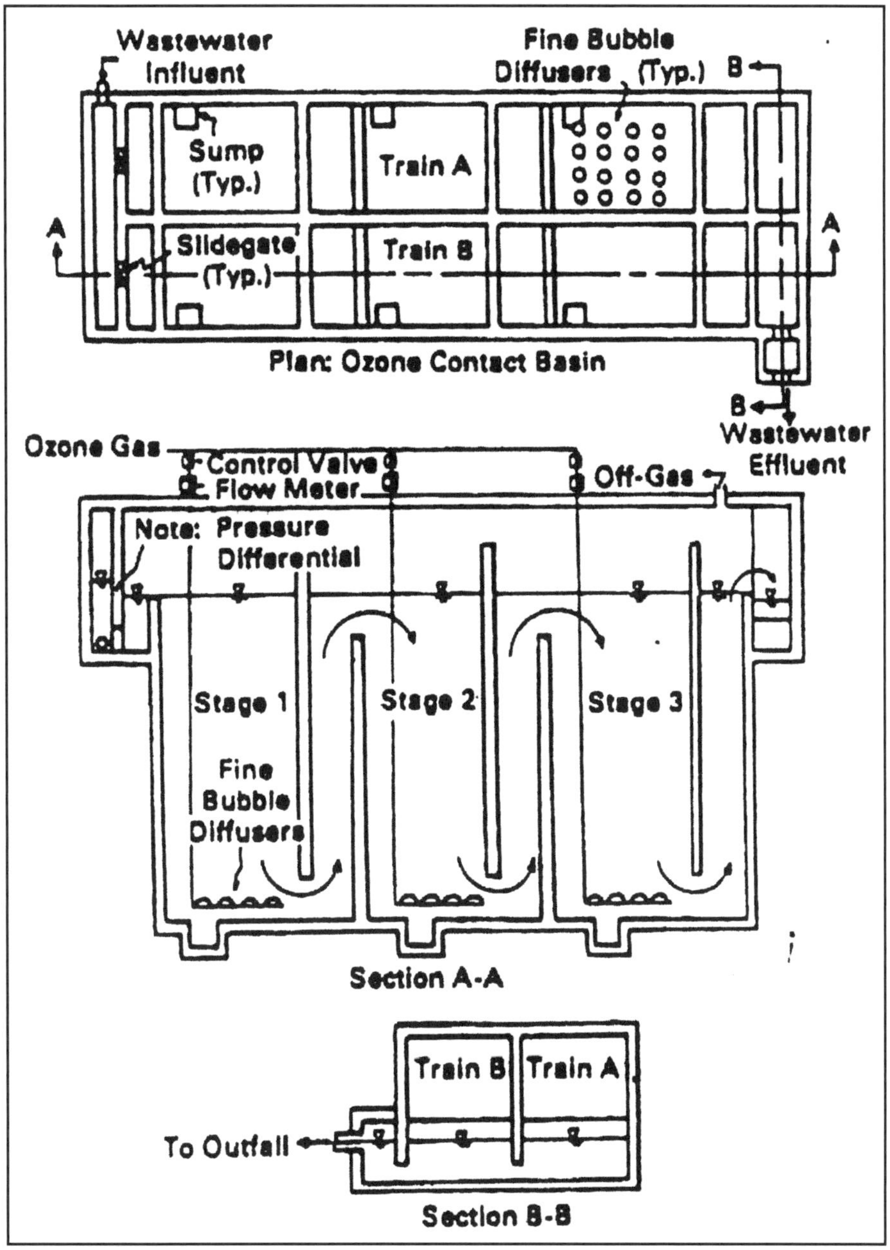

Figure 9.6. Schematic of a 3-stage, bubble diffuser ozone contact basin.

For MS2 inactivation, the pHs of water from Lake Audubon and Lake Sakakawea were both 8.4. The turbidities were 0.6 and 1.9 NTUs, respectively. Results indicated that within 1 minute, approximately 6 logs (99.9999 percent) of virus removal was observed. When no free chlorine contact time was provided (immediate addition of ammonia to form chloramines), a 2 log reduction (99 percent) was observed over 24 hours. Inactivation results were similar for both lakes' water.

For *Giardia muris* inactivation, several sets of water samples were tested. Lake Audubon samples had pHs varying from 7.9 to 8.2, and turbidities ranging from 0.4 to 6.4 NTUs. Samples from Lake Sakakawea had pHs of 7.9 to 8.1, and turbidities of 0.8 to 3 NTUs. Results indicated that with a 5 minute contact time and 3 hours of chloramine contact, approximately 3 logs (99.9 percent) removal was achieved. Without any free chlorine contact time, greater contact time was required to reach the same inactivation levels. Like the MS2 inactivation tests, both sets of lake water samples yielded similar results.

Surface water treatment requirements demand drinking water supplies be treated to a certain level of virus and *Giardia* removal. The required removals through the entire process are a 3 log (99.9 percent) removal of *Giardia lamblia* cysts and a 4 log reduction (99.99 percent) of enteric viruses (U.S. EPA, 1990). Because the required removal encompasses the entire plant, the rules credit the plant for different unit processes. Direct filtration is credited for a 2 log (99 percent) reduction in *Giardia lamblia* cysts and a 1 log (90 percent) reduction in viruses. This leaves a 1 log (90 percent) reduction in *Giardia lamblia* and a 3 log (99.9 percent) reduction in viruses that must be accomplished during disinfection.

For the Minot portion of the Northwest Area Water Supply Project, it is possible that water from Lake Sakakawea or Lake Audubon would not be filtered prior to disinfection because water would be piped to the Minot water treatment plant. If the surface water treatment rules for disinfection were satisfied before the water reached the divide between the Missouri and Hudson Bay watershed, the combination of disinfection in the pipeline and treatment at the Minot water plant would satisfy concerns over interbasin biota transfer.

In the conclusion of the chlorine challenge study, it was recommended that a free chlorine residual of 4 milligrams per liter with a 5-minute contact time before ammonia injection to form chloramines should be used to disinfect Garrison Diversion water before delivery across the divide. Based on their comparisons to surface water treatment requirements, a 3 log reduction in *Giardia muris* and a 4 log reduction in MS2 enteric virus could be achieved well before the water reached the watershed boundary in the pipeline. However, there would be considerably less margin for safety if the 5-minute free chlorine contact was not included.

In research conducted at the University of North Dakota, it was assumed that direct filtration would be performed prior to chlorine disinfection. The conceptual design calls for a 10-minute contact time and a free residual chlorine concentration of 0.5 milligrams per liter. This would correspond to a Ct value of 5 milligrams per liter * minute. Assuming a worst case scenario (a temperature of 5°C and a pH of 9), the Ct value required by the surface water treatment requirements for a 3 log reduction in viruses would be 4 milligrams per liter * minute. A 1 log reduction in *Giardia lamblia* cysts would require a Ct value of 95. Therefore, the design Ct value would not meet drinking water requirements. Several important implications can be inferred from these values. More treatment would be required before using this water for human consumption. This highlights the difference between the diversion planned in the Northwest Area Water Supply Project, where water will be treated to drinking water standards, and the eastern diversion, which is aimed strictly at biota transfer. In the eastern projects, prevention of biota transfer should treat the water to remove viruses and *Yersinia ruckeri*. In fact, the design was based on inactivation of *Yersinia ruckeri*. The design's Ct value would also be satisfactory to remove viruses because the design's Ct value of 5 exceeds the Ct value of 4 required for a 3 log virus reduction.

The key aspects of the conceptual chlorination design involve chlorine dose, chlorine contact basins, and a dechlorination system (Moretti et al. 1993; Moretti et al. 1996). Recall, that to achieve a free chlorine residual, the chlorine demand must be satisfied first. Chlorine demand was assumed to be about 1 milligram per liter so a total chlorine dose of 1.5 milligrams per liter should leave a free chlorine residual of 0.5 milligrams per liter. The dosage is important because it can be used to calculate the total quantity of chlorine that needs to be added for any given flow. To achieve a contact time of 10 minutes, the total tank volume required was calculated to be 13,125 cubic meters (465,000 cubic feet). The tank would have six baffles so a total of 7 compartments would be present in the tank. This configuration is shown in Figure 9.4.

Before entering the baffled contact chamber, it is also necessary to provide a mixing chamber for chlorine addition. Equipment required for this step will depend on the type of chemical added to establish a chlorine residual. Dechlorination components are very important to eliminate chlorine discharged back to the environment. The conceptual design proposes adding liquid sulfur dioxide. Sulfur dioxide will react nearly instantaneously with chlorine residual so that another contact chamber for chlorine removal is not necessary.

Ozone disinfection

While ozone has been widely applied in Europe for nearly a century, using ozone for disinfection of water and wastewater at large facilities in the United States has increased only over the last few years. The recent use of ozone in the United States is most likely due to increased awareness and concern over harmful byproducts formed from chlorine such as trihalomethanes. Although byproduct formation due to ozonation processes is usually considered negligible, there is mounting concern over ozone disinfection byproducts.

The foremost concern over ozone disinfection of drinking water supplies is that ozone does not leave a lasting residual in water. In water treatment plants, ozone is used as a disinfectant, and chlorine is subsequently added to maintain some disinfecting power in the distribution system. The residual problem is different in treating water for interbasin transfer. In the last section, a sulfur dioxide dechlorination system was discussed as a necessary feature of a chlorine disinfection facility for treating interbasin transfer water. The presence of chlorine residual or, for that matter, other disinfectant residuals is undesirable in these instances because of potentially detrimental effects in receiving waters. Therefore, ozone would seem to be especially well suited for biota removal in interbasin water transfers.

Chemically, ozone is a strong oxidant like chlorine. Ozone is formed by passing a pretreated gas that includes oxygen between two pieces of metal that have a very high voltage difference. Some of the energy formed due to the voltage difference is released to form ozone from oxygen molecules. Oxygen is stable as O_2. Ozone (O_3) is unstable and will rapidly decompose back to oxygen unless there are compounds to react with (i.e., there is an ozone demand). Like chlorine, ozone acts by disrupting the cell wall and protein systems of bacteria and viruses. Ozone has also been used for other purposes in water and wastewater treatment, some of which are the same as uses for chlorine. Although ozone can be an air pollutant and will irritate the human respiratory system, it is much safer to humans than chlorine gas, even at high concentrations. If there is an ozone leak in a plant, the staff can smell the ozone before the levels are dangerous. This is another distinct advantage of ozone over chlorine. Ozone disinfection has many benefits over chlorine, especially when no disinfectant residual is required. However, the use of ozone requires a significant investment in additional equipment and electrical power.

Ozone generation depends on a gas flow with oxygen. The gas is fed into ozone generators where it passes between two electrodes. The electrodes are charged to give a very high electrical potential between the electrodes. This electric potential (voltage) results in current flow from one electrode to the other, separating the two atoms of molecular oxygen into atoms and the atoms can reassociate to form the three-atom ozone molecule. The same phenomenon occurs in the atmosphere during electrical storms. Ozone is the fresh odor encountered after electrical storms. Using ozone will significantly increase electrical requirements of the facility.

The feed-gas characteristics are very important to ozone generation and treatment plant operation. As one would expect, if the feed gas has a high concentration of oxygen, the ozone concentration in the gas coming from the ozone generators will be much higher than if air is used. The presence of moisture and dust particles in the feed gas is of utmost importance for the efficiency of the ozone generators. Dust particles will coat the electrodes, reducing the ozone generation efficiency. Likewise, the presence of moisture will damage the ozone generators; thus, a clean, dry feed gas is required. Fine filters, perhaps in a series, are usually adequate to remove particulates even if particles are smaller than 1 micron wide (0.00004 inches). Moisture is usually removed by some sort of dessicant which absorbs the moisture from the air. The feed gas must be dried to a dewpoint of around -50C. There are some advantages to using liquid oxygen rather than an air feed system. Usually liquid oxygen is low in moisture and free from dust particles. However, because of the extreme explosion hazards and potential cost involved, many ozone systems use air feed systems.

After the feed gas passes the ozone generators, it enters the ozone contact chamber. Here ozone in the gas stream can disinfect the water. Because ozone is applied in a gaseous form, the ozone contact chambers are different than the chlorine contact chambers. Chlorine contact chambers are usually open to the atmosphere with vertical baffles to route water horizontally through the basin. Ozone contact chambers are sealed to the atmosphere, and water typically moves vertically up and down the compartments. The primary reason for this difference is that ozone is applied as a gas to the water.

Ozone has to dissolve to react with contaminants in water. Usually in the first compartment of an ozone contact chamber, water enters at the top and flows down before entering the second compartment. The ozone bearing gas enters the first compartment at the bottom through a system of diffusers. Diffusers make small bubbles of the ozone-bearing gas that pass up through the downflowing water. Some of the ozone (and other gases present in the feed gas) dissolve in the water. The gases that do not dissolve must be vented at the top of the compartment. Therefore, not all of the ozone applied to a contact chamber will end up in the water. This introduces the concept of transfer efficiency. The transfer efficiency is the percentage of the ozone dissolved in the water out of the total generated ozone. This can be determined by measuring the flow rates of gas into and out of the chamber and measuring the respective concentration of ozone. The transfer efficiency assumed in the conceptual design of the ozone disinfection facility was 92 percent (Moretti et al. 1993). The transfer efficiency in the pilot experiments conducted with a small contact chamber and ozone generator was closer to 50 or 60 percent (Richard et al. 1996).

The lower the transfer efficiency, the less ozone is dissolved in the water, and the more ozone is lost with the vented off-gas. Several options are available for the lost ozone. The off-gas cannot legally be discharged directly to the atmosphere. Ozone is a pollutant in the lower atmosphere and can contribute to smog formation, so off-gas has to be treated before release. This is usually done by heating the off-gas or passing it through a metal catalyst. Both types of ozone destruction units are common. Once ozone has been removed, the off-gas can be vented to the atmosphere or used again as feed gas. This is possible because only a very small fraction of the oxygen in the feed gas is actually converted to ozone. Furthermore, ozone decays to oxygen so some of the ozone generated will decompose back to molecular oxygen that will be present in the off-gas. Another option for the feed gas is to return at least some of it to the contact chamber. Often plants will be designed so that there are diffusers in all the compartments of the contact chamber. Since there is ozone in the off-gas it can be applied to another compartment providing more treatment.

There are several design features to improve transfer efficiency and minimize ozone generation requirements. The first relates to the size of bubbles formed from the diffusers. The smaller the bubbles, the greater the surface area available for gases to dissolve into the water; therefore small bubbles are desirable. Usually a bubble size of 2-3 millimeters is recommended (Cruise, 1993). Other factors like turbulence and contact time also influence the transfer efficiency. Generally, there will be a fair amount of turbulence in compartment(s) where ozone is applied due to the action of the diffusers. The first compartment of the ozone contact chamber is almost invariably operated in a downflow mode. This maximizes the contact time available for the ozone to dissolve. To maximize the solubility of ozone, the contact chambers can be operated at a small positive pressure. The solubility of gases is also governed by temperature. The solubility of gases increases in cold water and decreases in warm waters.

Ozone that dissolves in water is available to satisfy the ozone demand of water and to establish an ozone residual. Like chlorine, ozone will react with numerous organics and reduced inorganics depleting the pool of ozone available to disinfect the water. The strong oxidizing ability of ozone has been used to help eliminate metals like iron and manganese, control taste and odor problems, and destroy pesticides in drinking water. Therefore, of the ozone generated, some is lost due to the transfer efficiency concept, some is used to satisfy the ozone demand of the water, and the rest will establish an ozone residual.

Once the residual is established, disinfection can occur. Ozone disinfection of drinking water is governed by the Ct concept. Ozone is very unstable, causing the ozone residual concentration to decrease as the water flows through the contact chamber. Research conducted at the University of North Dakota (Cruise 1993; Moretti et al. 1993) brought the ozone residual to a set concentration and maintained this concentration for a set time frame. The Ct value was determined by multiplying residual concentration and contact time. Research conducted for the Northwest Area Water Supply project (Houston Engineering et al. 1996) and at North Dakota State University (Richard et al. 1996) allowed the residual to dissipate over a period of time. In these cases, a mathematical equation was used to determine the Ct value because the residual changed over the contact time.

Ozone disinfection will result in byproducts. Foremost among these are bromate and aldehyde compounds like acetylaldehyde and formaldehyde. The Northwest Area Water Supply Project levels of these byproducts, among others, formed during some of the ozonation tests (Houston Engineering et al. 1996). In general, the concentration of ozone byproducts was higher for Lake Audubon, which is consistent with the higher organic concentration and disinfectant demand. None of the samples tested contained measurable quantities of bromate, the most strictly regulated ozone disinfection byproduct.

Although the main goal of the Northwest Area Water Supply research was to determine the feasibility of using chlorine to disinfect Lake Sakakawea water before the water is transferred to the Souris basin, several sets of experiments were conducted with ozone. Ozone demand for Lake Sakakawea water was greater than for Lake Audubon, consistent with the higher level of organics and the corresponding higher chlorine demand. These also correlate to higher formation of both chlorine and ozone disinfection byproducts. For both waters, the ozone demand was between 2.4 and 3.1 milligrams per liter at 20°C. These values were slightly lower at 4°C, indicating that at lower temperatures, ozone decays at a slower rate.

Giardia muris cysts were challenged in inactivation tests. Results indicated that an initial ozone residual of 0.3 milligrams per liter achieved a 3 log reduction in cysts at 4°C. The action of ozone on viruses is greater than on cysts, so no MS2 virus inactivation tests were performed. *Giardia* inactivation continued even after ozone residual had completely dissipated. Normally the Ct value is determined when the ozone has dissipated during an inactivation test. It was speculated that byproducts of ozone decay continued to inactivate the cysts even after the ozone residual was zero. Because of this phenomenon, the Ct concept was not applied to the experimental results.

The first step in evaluating the ozone disinfection of Lake Sakakawea water at the University of North Dakota was to determine the ozone demand by adding ozone to water samples and trapping the off-gas. The difference between the ozone applied and the ozone not measured as a residual or trapped in off-gas was the ozone demand. Tests were performed at 19, 14, and 6°C with several water samples. The different temperature waters were from July, October, and February. Water samples tested included both unfiltered raw water samples and samples filtered after being treated with alum and/or a polymer to enhance filter performance. Results indicated that the ozone demand did not vary greatly with the level of pretreatment, but the demands were higher at warmer temperatures. The higher demands with warmer temperatures was consistent with results from the Northwest Area Water Supply experiments. Even when raw water with turbidities of up to 19 NTUs were tested, their ozone demand was only about 2.5 milligrams per liter. This corresponded well with demands observed in the Northwest Area Water Supply experiments when raw waters had a demand between 2.4 and 3.1 milligrams per liter. For the filtered water samples, the demands were usually slightly lower. This has great implications if ozone would be used for disinfecting Garrison Diversion water. The very slight difference in ozone demands between raw and filtered water means that unfiltered water could be treated without significantly increasing the ozone demand and, in turn, ozone generation requirements. Based on the experiments, an ozone demand of 2.0 milligrams was assumed for the conceptual design.

The inactivation results showed that naturally occurring lake bacteria that passed through the direct filtration process were completely inactivated with an ozone residual of 0.09 milligrams per liter with a contact time of 5 minutes at 14°C. This corresponded to a Ct value of 0.45 milligrams per liter * min. One run with raw water (turbidity = 19 NTU) required a residual of 1.44 milligrams per liter to achieve a 99.9 percent inactivation at 14°C. This corresponds to a Ct value of 7.2 milligrams per liter * min. This demonstrates that, while the turbidity of raw water may not cause a dramatic increase in ozone demand, it does dramatically reduce the disinfection efficiency of ozone. This is consistent with the shielding concept where microorganisms in the presence of turbidity are somehow protected against disinfectants. This also supports direct filtration of Garrison Diversion water prior to disinfection.

Inactivation tests with *Y. ruckeri* indicated that this organism was relatively sensitive to ozone. Tests were conducted at 2 and 20°C, although there was little difference in the results between the two temperatures. An ozone residual of 0.13 milligrams per liter inactivated 99.9 percent of *Y. ruckeri* within 3 minutes (Ct = 0.39 milligrams per liter * minute), and a residual of 0.38 milligrams per liter for 3 minutes (Ct = 1.14 milligrams per liter * minute) resulted in a complete inactivation.

Therefore, a residual of 0.2 milligrams per liter and a contact time of 1 minute was established for the conceptual design. Based on laboratory results, it was concluded that at the design conditions, (Ct =

0.5 milligrams per liter * minute) greater than 99.9 percent of *Y. ruckeri* and total bacteria in the filtered lake water would be inactivated. Assuming a worst case scenario for disinfection of filtered water from Lake Sakakawea (a temperature of about 1°C), a Ct value of 0.97 is required to treat Garrison Diversion water that has been treated by direct filtration to achieve a 1 log reduction in *Giardia* cysts to satisfy drinking water requirements. Under the same conditions, a 3 log reduction in viruses would require a Ct value of 1.4 milligrams per liter * minute. Like the conceptual design for a chlorination system, the ozone design would only treat water based on inactivation of *Y. ruckeri* and other relatively sensitive microorganisms rather than provide drinking water level inactivation.

The conceptual design calls for air for the feed gas and an operational Ct of 0.5 milligrams per liter * minute. Like the chlorination and UV system, the scale of the ozone contact chamber is impressive. The design calls for a three compartment contact basin with stone diffusers to bubble the ozone into the contact chamber. Allowing for 1.8 meters (six feet) of freeboard and a liquid depth of 4.3 meters (14 feet), the basin would be 6.1 meters (20 ft) tall. To treat the design flow with the Ct constraints determining contact time and ozone generation capacity, the basin would have to be 28.7 meters (94 feet) wide and 86 meters (282 feet) long. This would result in a total volume of over 15,000 cubic meters (about 530,000 cubic feet).

Research conducted in two phases at North Dakota State University included some other aspects of ozone disinfection of Garrison Diversion water. In the first phase, batch experiments, similar to experiments conducted with the Northwest Area Water Supply Project, were conducted to evaluate ozone demand and residual decay characteristics of water from Lake Sakakawea. A continuous flow pilot plant was developed from these results. In phase two, the pilot scale plexiglass ozone contact chamber was used to evaluate the relationship among the ozone demand, residual, and decay while water was passing through the contact chamber. Several experiments were also conducted to further elucidate the activity of ozone on *Y. ruckeri* .

In the experiments with the pilot contact chamber, water samples from Lake Sakakawea were brought to Fargo, North Dakota, and pumped through the contact chamber. An air feed was used to generate ozone, and ozone was applied to the first contact chamber compartment. Water entered the contact chamber and flowed down in the first compartment. This maximized contact time between applied ozone and downflowing water. Different ozone dosages were applied in different runs. Eight sample ports allowed the ozone residual at various points in the contact chamber to be tested. The ozone concentration in the off-gas was also determined. This experimental set-up allowed ozone to dissipate naturally in a chamber with hydraulic characteristics similar to a full-size contact chamber. Critical parameters, like the demand and decay in actual Garrison Diversion water, were determined in an actual contact chamber; albeit small scale. These would be vital design parameters if an ozone treatment system were ever used. Continued testing of Lake Sakakawea water during this research also offered more opportunity to determine applicability of Garrison Diversion water over another year with potentially different water quality parameters.

A range of microbiological tests was also conducted. Foremost among these were over 60 inactivation experiments where pure cultures of *Y. ruckeri* were inactivated at different temperatures and with different initial concentrations of microorganisms. Other tests confirmed the difficulty of evaluating the removal of naturally occurring lake bacteria, including *Y. ruckeri*, with conventional methods. Another set of experiments confirmed that *Y. ruckeri* was capable of persisting very well in water from Lake Sakakawea, further demonstrating the need to eliminate the organism before transferring water across the drainage divide.

During phase one, batches of unfiltered water from Lake Sakakawea were dosed with ozone. Ozone was applied to the batch until a given dose had been applied. Then the ozone application was stopped, and the residual was allowed to dissipate. The residual concentration was monitored with time to determine decay characteristics of the ozonated water. The amount of ozone actually applied to the water sample was determined by trapping the ozone in off gas. Therefore, the total amount of ozone applied was determined (i.e., transfer efficiency). Results showed that of the ozone transferred to unfiltered Lake Sakakawea water, about 80 percent was used to satisfy the demand of the water, and about 20 percent ended up as a measurable residual.

Decay curves were required to determine a relationship between the initial concentration and the resulting Ct value. There is little control over contact time in a continuous flow system because, once ozone is applied, it will decay. Therefore, contact time is determined by the initial residual and the residual decay characteristics. The ozone decay curve shown in Figure 9.3 is a typical curve. The curve can be modeled with a linear or exponential mathematical equation. Both types of mathematical equations were used to describe the experimental data, and it was concluded that both types of equations could be used to describe the decay in Lake Sakakawea water. Once the mathematical equations of ozone decay were determined, the Ct value for a given initial residual was determined. With all the data, a relationship was finally developed among the transferred dosage, the residual achieved, and the corresponding Ct value. This relationship could determine the Ct value achieved for a given dose of ozone transferred to a batch of water from Lake Sakakawea. The usefulness of such a relationship is apparent if one knows the transfer efficiency of a system and the concentration of ozone in the gas applied. If these parameters are known, the operators of an ozone disinfection facility can determine how much ozone to apply to maintain a given Ct value, given the relationship between the transferred dose and Ct value. Furthermore, they can check the initial residual to determine if demand is changing and adjust the ozone application rate if necessary.

The pilot contact chamber was designed and constructed from phase one results (the transferred dose - Ct relationship). The goal of the experiments conducted with continuous flow were very similar to phase one. Ozone was applied to the system; the transfer efficiency, demand, and initial residual and Ct values were determined. Again relationships among the transferred dose, initial residual, and corresponding Ct value were developed. This provided the key operational parameters of a continuous flow ozone disinfection system.

Three water samples were tested during the continuous flow experiments. They were an unfiltered sample from deep in the lake (July 1995), an unfiltered sample from the surface of the lake (August 1995), and a pressure-filtered sample from the bottom of the lake (August 1995). For the unfiltered samples, about 90 percent of the transferred dosage was used to satisfy the ozone demand, and 10 percent ended up as a measurable residual. In pressure-filtered water, on the other hand, only about 70 percent of the transferred dosage was needed to satisfy ozone demand, and 30 percent ended up as residual. Based on limited results, filtering Garrison Diversion water significantly reduced the demand of the water. For samples tested at the University of North Dakota, it did not appear that filtering would have that big an impact on ozone demand. The difference, if there is a significant one, might be attributed to something as simple as slightly different water characteristics. In turn, it was easier to achieve a target Ct value with low doses of ozone in the filtered water.

Spurred by the demonstration that ozone disinfection of Lake Sakakawea was a feasible method to control *Yersinia ruckeri* and, potentially, the IHNV virus, an exhaustive literature review of this type of ozone application and of *Yersinia ruckeri* was conducted at North Dakota State University (Richard et al. 1992; Woodbury et al. 1992; Richard et al. 1996). This led to a much better understanding of this type of ozone application. Several key points merit additional attention. First, if ozone was used to disinfect water in a direct filtration plant, ozone would be beneficial in many areas besides disinfection. Ozone enhances removal of organics and suspended solids during direct filtration. This would enhance removal of both types of compounds in direct filtration. The lower suspended solids (turbidity) in filtered water would enhance disinfection efficiency. This was demonstrated in ozone inactivation tests conducted at the University of North Dakota where it was not possible to completely eliminate lake bacteria in unfiltered water with a high turbidity. The enhanced removal of organics during direct filtration might also reduce the potential to form ozone byproducts like bromate and various aldehydes. Ozone is also effective in destroying potentially harmful compounds like pesticides that may be present in Garrison Diversion water.

The use of ozone also has to be carefully controlled. At first glance, it would appear that because ozone rapidly decays into molecular oxygen, ozone treatment would be beneficial in providing excess oxygen into the receiving water. However, fish are extremely sensitive to not only ozone, but very high concentrations of oxygen. If ozone were used, residual ozone and excess oxygen would have to be dissipated before allowing disinfected water to enter receiving waters. Another potentially detrimental impact relates to stream purification. Ozone reacts with some organics making them more biodegradable. Microorganisms could rapidly degrade these altered organics and deplete oxygen available for other organisms like fish. Water from Lake Sakakawea is very low in dissolved organics and, therefore, should not cause concerns.

There are many different types of *Yersinia ruckeri*. Some strains do not appear to be very dangerous to fish. Other, virulent strains may show different responses to ozone inactivation, although this has not been reported. Another important aspect of an ozone disinfection study should be ozone regrowth. Organisms exhibiting this property would not grow during normal enumeration methods, but may grow after an extended period of time, much like organisms capable of repairing UV-damaged DNA. Previous work showed no signs of ozone regrowth with *Yersinia ruckeri* (Wedemeyer and Nelson 1977). Current work did not clearly demonstrate regrowth but could not rule out the possibility of this phenomenon occurring.

Ozone inactivation tests were conducted at approximately 10 and 20°C. The majority of tests were conducted in a phosphate-buffered saline solution suitable for not harming microorganisms because tests in lake water exert some ozone demand. Microorganisms grown on petri plates prior to disinfection were washed with saline to prevent any organics from the media, or excreted during growth of organisms, from exerting an ozone demand. The experiments were conducted at initial organism concentrations varying from approximately 9,000 to 40,000,000 organisms per milliliter. Tests with lake water spiked with washed microorganisms were also conducted to verify the applicability of saline inactivation. Results indicated that *Yersinia ruckeri* could be readily inactivated if the surface water treatment requirements for a 3 log of reduction viruses were followed .

Monitoring of a disinfection system treating water to remove foreign biota also deserves attention. In the North Dakota State University research, a working assumption was that the treatment level should be similar to drinking water requirements (i.e., using drinking water Ct values). Water treatment plants and wastewater treatment plants rarely test for a specific pathogen. They test for groups of microorganisms that, if present, indicate that treatment efficiencies are not adequate and that pathogens could be present. If only a handful of organisms were being monitored, as in the case of the Garrison Diversion Unit, it would seem that testing for these organisms would be an easy extension of current monitoring techniques. However, this is far from true when considering normal laboratory techniques. *Yersinia ruckeri,* for example, when growing outside of a host in natural waters will not be easily recovered on the same growth media used to isolate it from infected fish. Although the microorganism may not grow on traditional media, it can still infect fish. Once an isolate characteristic of *Y. ruckeri* is detected, after perhaps 48 hours of incubation, this isolate must be tested with other biochemical tests in traditional diagnostic techniques. There are rapid serological techniques to determine if the isolate is, in fact, the pathogen of interest, but this isolate still has to be grown first. This is also not a normal technique for water or wastewater treatment plants, so specialized personnel with specific training may be required. Even more importantly, if *Yersinia ruckeri* were detected in effluent, it would likely be two days after the fact. Therefore, it is obvious that if a filtration/disinfection system is constructed, questions remain about how to monitor performance. Like the fish screen problem of nearly 20 years ago, many questions still have to be answered.

References

Bullock, G.L., and H. M. Stuckey. 1977. Ultraviolet Treatment of Water for Destruction of Five Gram-Negative Bacteria Pathogenic to Fishes. *Journal of the Fisheries Research Board of Canada* 34:1244-1249.

Clambey, G.K., H.L. Holloway, J.B. Owen, and J.J. Peterka. 1983. Potential Transfer of Aquatic Biota Between Drainage Systems Having no Natural Flow Connection. Fargo: N. Dak.: Tri-College University Center for Environmental Studies.

Cruise, T.L. 1993. *Use of Ozone to Remove Fish Pathogens from Garrison Diversion Water.* M.S. thesis, University of North Dakota, Grand Forks.

Harza Engineering. 1977. *A Review of the International Garrison Study Board Report.* Report to the Garrison Diversion Conservancy District.

Hefta, M.J. 1990. *Direct Filtration of Garrison Diversion Water.* M.S. thesis, University of North Dakota, Grand Forks.

Houston Engineering, American Engineering, and Montgomery Watson. 1996. Northwest Area Water Supply Project Chloramine Challenge Study. Prepared for the North Dakota State Water Commission and the Garrison Diversion Conservancy District.

International Garrison Diversion Study Board. 1976. *Report to the International Joint Commission.*

International Joint Commission. 1977. *Transboundary Implications of the Garrison Diversion Unit.*

Jensen, K. 1991. *A Review of Interbasin Water Transfers.* M.S. thesis, North Dakota State University, Fargo.

Johnson, D.L. 1975. Hydraulic Model Studies of a Fish Screen Structure for the McClusky Canal. U.S. Bureau of Reclamation, Report REC-ERC-75-6.

Kopchynski, D.M. 1991. *Disinfection Effectiveness of Chlorine, Ozone, and UV Light on Selected Fish Pathogens.* Special Topics Paper, Department of Civil Engineering, University of North Dakota, Grand Forks.

Ludwig, H.R., and J.A. Leitch. 1995. *Pathways for Aquatic Biota Transfer Between Watershed.* North Dakota Water Resources Research Institute, North Dakota State University, Fargo.

Metcalf and Eddy, Inc. 1979. *Wastewater Engineering.* New York: McGraw-Hill.

Moretti, C.J., and D.M. Kopchynsk. 1992. Selection of filtration media for optimal removal of biota. *Proceedings of the North Dakota Water Quality Symposium*, Mar. 25-26, North Dakota State University Extension Service, North Dakota State University, Fargo, pp. 198-204.

Moretti, C.J., D.M. Kopchynski, and T.L. Cruise. 1993. *Evaluation of Direct Filtration and Disinfection for Prevention of Biota Transfer into the Hudson Bay Drainage.* Water Resources Research Institute, North Dakota State University, Fargo.

Moretti, C.J., D.M. Kopchynski, and T.L. Cruise. 1996. Controlling microbial biota transfer in the Garrison Diversion Unit, *ASCE Journal of Water Resources Planning and Management* 122:197-204.

Padmanabhan, G., K. Jensen, and J.A. Leitch. 1990. A review of interbasin water transfers with specific attention to biota. *Proceedings of the Symposium on International and Transboundary Water Resource Issues*, American Water Resources Association, pp. 93-99.

Richard, O.D., L.H. Woodbury, and G. Padmanabhan. 1992. *Ozonation as a Water Treatment Method for Prevention of Interbasin Transfer of Biota: A Bibliographic Review.* North Dakota Water Resources Research Institute, North Dakota State University, Fargo.

Richard, O.D., R.A. Zimmerman, and K.E. Rosvold. 1996. *Laboratory Study of Feasibility of Ozonation to Control Fish Pathogens.* Water Resources Research Institute, North Dakota State University, Fargo.

Sako, H., and M. Sorimachi. 1985. Susceptibility of fish pathogenic viruses, bacteria, and fungus to ultraviolet irradiation and the disinfectant effect of UV-ozone water sterilizer on the pathogens in water. *Bulletin of the National Research Institute of Aquaculture* 198:51-58.

Sayler, R.D. 1990. Fish transfer between the Missouri River and Hudson Bay basins: A status report and analysis of biota transfer issues. *1990 Interbasin Biota Transfer Study Program Proceedings.* North Dakota Water Quality Symposium, Mar. 20-21, Water Resources Research Institute, North Dakota State University, Fargo, pp. 25-40.

Turner, C.D. 1988. Removal of fish pathogens from inter-basin transfer water. *Proceedings of the 40th Annual Convention.* Western Canada Water and Wastewater Association, pp. 71-77.

Turner, C.D., and M.J. Hefta. 1989. Evaluation of direct filtration for prevention of biota transfer into the Hudson Bay drainage. *Proceedings of the North Dakota Water Quality Symposium*, Mar 20-21, North Dakota State University Extension Service, North Dakota State University, Fargo, North Dakota.

U.S. Bureau of Reclamation. 1982. *McClusky Canal Fish Screening Facility.* Report presented to Regional Director, Billings, Mont.

U.S. Department of the Interior. 1982. *McClusky Canal Fish Screen Development and Verification.*

U.S. Environmental Protection Agency. 1990. *Guidance Manual for Compliance with the Filtration and Disinfection Requirements for Public Water Systems Using Surface Water Sources.* Washington, D.C.: Environmental Protection Agency.

Viessman, W., Jr., and M.J. Hammer. 1985. Water Supply and Pollution Control. New York: Harper and Row.

Water Pollution Control Federation. 1984. *Wastewater Disinfecion; A State-of-the-Art Report.* Water Pollution Control Federation, Alexandria, Virginia.

Water Pollution Control Federation. 1986. *Wastewater Disinfection.* Water Pollution Control Federation, Alexandria, Virginia.

Wedemeyer, G.A., and N.C. Nelson. 1977. Survival of two bacterial fish pathogens *(Aeromonas salmonicida)*, and the enteric redmouth bacterium in ozonated, chlorinated, and untreated waters. *Journal of the Fisheries Research Board of Canada* 34: 429-432.

Wedemeyer, G.A., N.C. Nelson, and C.A. Smith. 1978. Survival of the salmonid viruses Infectious Hematopoietic Necrosis (IHNV) and Infectious Pancreatic Necrosis (IPNV) in ozonated, chlorinated, and untreated Waters. *Journal of the Fisheries Research Board of Canada* 35:875-879.

Woodbury, L.H., O.D. Richard, G. Padmanabhan, and M. McLaughlin. 1992. Ozone as a water treatment method for prevention of interbasin transfer of biota. *Proceedings of the North Dakota Water Quality Symposium*, Mar. 25-26, North Dakota State University Extension Service, North Dakota State University, Fargo, pp. 205-216.

CHAPTER 10
Summary, conclusions, and implications

Jay A. Leitch

As evident from the two forewords, there are at least two philosophically divergent views of interbasin transfer of aquatic biota as it relates to a specific project in North Dakota. Each of these views evolved over a long period of time, building upon numerous sources of information from layperson hunches to hard science. Each view is likewise internally consistent, logically legitimate, and understandable by objective observers. These divergent views, however, have contributed to the controversy surrounding the Garrison Diversion Project. As a direct result of these unresolved differences, the Project has neither moved forward nor been dropped from consideration. This deadlock prompted the U.S. Congress and leaders in North Dakota to implement a "scientific" study of aquatic biota transfer.

The scientific study began in 1988 with support from the Garrison Diversion Conservancy District, the North Dakota State Water Commission, and the US Bureau of Reclamation, three sponsors with an obvious interest in the Garrison Diversion Project. While project sponsors may have hoped for quick, decisive answers supportive of their biases, a strong project director and an objective study process were put in place to ensure objectivity as well as scientific credibility. After several years, several million dollars, and several focused research projects, the study reached a point of diminishing scientific returns. In other words, science will never provide "the answer" and, for policy making purposes, more and more science adds less and less to the debate.

Role of science in policy

Policy makers rely on a variety of information and advisors when assessing alternatives. Among that variety is scientists and science. While the hope may sometimes be that science will dictate an alternative, it cannot; it can only more clearly illustrate the consequences of alternatives. In the end, social decision making is more than science. Such is certainly the case in this study.

Science can describe and help systematically evaluate the alternatives, but

- science alone cannot resolve interbasin biota transfer issues,
- science can reduce the uncertainty surrounding technical decisions,
- science cannot eliminate uncertainty,
- science can narrow the range of plausible choices, and
- science cannot deliver "truth," it can only reject untruth.

Thus, the role of science is not to provide answers, but rather to narrow the range of choices and more clearly inform policy makers of the consequences of plausible choices.

History of Garrison Diversion

The Garrison Diversion Project has had a long and sometimes stormy history. Initially, a part of the Pick-Sloan Plan in the mid-Forties, the Garrison Diversion Conservancy District was established in the Sixties to complete the irrigation component of the plan. From the start, the District Project was opposed by U.S. and Canadian environmental groups and others. The Project was reformulated by a U.S. Federal committee in 1985, which also mandated an international study group be developed to investigate the biota transfer issues. That was the start of this study.

Interbasin water transfers

Interbasin transfers of water are not new phenomena. The frequency, volume, and distance of transfers have increased over time as human populations demand more water in water-short regions. However, most of the science examines only engineering topics. Only a few case studies on aquatic biota transfer can be found in the scientific literature. Recently, concern over aquatic biota transfer has slowed the number and extent of such transfers—this study being a case in point—for these reasons:

- more careful examination (science) is needed of the issue;
- such examination requires broad-based, multidisciplinary approaches; and
- efficient solutions need to be identified for public acceptance.

Multiple pathways

There are many pathways for waterborne biota to move from one watershed to another. Bait bucket transfer and high water, for example, are two pathways that "connect" the Red River basin with the Missouri River basin. The record flood of 1997 in the Red River of the North Valley provided evidence that "natural" connections between the two watersheds occur. Exotics like the zebra mussel and Eurasian milfoil are thousands of times more likely to be transported by anglers than by water diversions. It does not make economic sense to spend a lot of time and effort to plug a tiny hole when there are several large holes, like bait bucket transfer or aquaculture, that are ignored. Economics would imply that effort be directed toward the area where the greatest reduction in transfer potential could be achieved.

Transport regulations, public education, regular inspection of boats and/or livewells, restriction of baitfish species, and more careful inspection of aquarium fish are feasible ways to control transport and release of bait and aquarium fish. Structural measures could be used to reduce the risks inherent with high-water years, but not to eliminate them.

Consequences of aquatic biota transfer

The potential consequences of the interbasin transfer of non-indigenous aquatic species vary from highly positive to highly negative. Consequences range from ecological to financial. The actual, ultimate consequence may be dynamic, and it may vary from all predictions because of exogenous factors, such as climate change. The science is incomplete regarding currently identified species of concern to Garrison Diversion Project interests. As well, there are yet to be identified species that may become concerns; obviously, little can be said about these.

Potential consequences are not clearly spelled out in this text, primarily due to the large number of variables and the high degree of uncertainty. All decisions involve uncertainty, but policy makers have different comfort levels incorporating those into their choices. Part of the discomfort is the expected balance between gains and losses. In short, all choices, including no action, have consequences, real or perceived. The policy goal is to minimize the negative and to maximize the positive consequences. Shying away from the negative may not be the optimal choice for society.

Aquatic species distribution

It is no small task to attempt an inventory of aquatic species in such a large watershed that contains so much surface water. In fact, a complete inventory of aquatic species is impractical for such large watersheds. Having a species on such an inventory indicates its presence at that place and point in time. Multiple "sightings" add weight to the argument that the species maintains a viable population in the watershed. However, not finding a species is not a certain indication that it is not there.

Scientists working on projects within the overall study identified several aquatic species, such as the rainbow smelt, in places where they had not been reported previously. This is expected, although the number of new species found in a watershed is likely to diminish over time barring large scale introductions of species. The study produced a much more complete description of the aquatic biota community than previously existed, and strong evidence to suggest species distribution is not static.

Aquatic species invasions

There are a number of explanations about how, why, and what impact invasive species may have in a watershed. Some species that may be thought to "invade" a new watershed may, in fact, be returning to a former niche after being displaced by a major event, such as global cooling (i.e., glaciers). Successful invaders may be merely repopulating an area previously occupied by them. Such invasions would have little or no consequences on the watershed's biodiversity.

Some specific findings about species invasions coming from this study include the following:

- invasions can happen in any direction (south as well as north);
- fish colonizing the Hudson Bay drainage from the Mississippi and perhaps Missouri headwaters have evolved in the presence of those species already present in the Hudson Bay drainage (the "coming home" hypothesis);
- fish colonizing the Hudson Bay drainage from the Mississippi and perhaps Missouri headwaters may just recreate the same niche as occupied in southern tributaries, given that similar prey items and habitat are available for exploitation, with little if any interaction with native species in the receiving watershed;
- aquarium species likely to colonize the Hudson Bay drainage are temperate European and North American species that could tolerate the prolonged winters;
- exotic introductions from other continents or distant North American drainage basins have not evolved with species of central North America and will have to create their own niche in central North American waterways, with higher probability for significant interactions with North American native species; and
- of the species known to be introduced to the Hudson Bay system to date, the carp and the smelt probably have the greatest impact to both aquatic ecosystems and economic activity.

Parasites and pathogens

Some parasites and pathogens not previously known to exist in the Hudson Bay drainage were found there during this study. In fact, some of those identified by Garrison Diversion opponents were found in the Hudson Bay drainage. Just because a species is not documented to exist in a watershed is not proof it is not there—we sometimes have to look harder to find it. Some specific conclusions about parasites and pathogens stemming from this study include the following:

- The discovery of two new species of parasites from lake sturgeon indicates the lack of complete information on biodiversity of the watershed.
- Forty new records of parasites in fish in Canadian waters also indicate a lack of complete information on watershed biodiversity.
- The changing mix of fish hosts harboring parasites which impact commercial fishing may not be related to aquatic biota transfer, but to other factors such as global warming or bait bucket transfer.
- Future studies dealing with biota transfer must be multidisciplinary, with the vision to predict natural processes that may otherwise confound biota transfer studies as background noise.

Water treatment

Many water treatment mechanisms have been suggested as ways to prevent the transfer of unwanted biota via interbasin transfers of water. Among those suggested are screens, chemicals, and biocontrols. The technology exists, using one or a combination of techniques, to treat Missouri River water well beyond any reasonable (i.e., efficient) risk management level. Ozone treatment was found to result in significant risk reduction in this study.

Efficient risk management involves reducing a risk to a level to just where the added costs exceed the added benefits of reduction. Risk reduction beyond this point is not an efficient use of society's resources. However, the level of acceptable or reasonable risk varies with perspective and whether the risk is voluntary or involuntary. Some who would face an involuntary risk as a result of biota transfer argue for zero or near-zero risk, while others argue for a risk more in line with accepted drinking water standards. Risk communication is an area where scientists and policy makers need to work together more closely.

So what?

This study set out to answer five general questions:

- What are the current (1997) concerns regarding the project's potential for aquatic biota transfer? **We found that Canadians were concerned about several specific fish species and the potential damage they might cause to fish communities and economic activity. They were also concerned about "as yet unidentified" species.**
- What is the present distribution of species in the areas of concern? **The distribution is dynamic, with species showing up in new places on a regular basis.**
- Why are not the fish species of concern already in certain waters? **Some species are not in certain waters either because they have not arrived there yet (they will eventually) or they cannot survive the environment (unless it changes, e.g., climate change).**
- What impacts would nonindigenous species have in certain waters? **The impact varies from positive to negative and from almost none if it was a formerly displaced species to substantial impact on biodiversity in the case of some invasive exotic species. There are too many potential species to know with much certainty what each would do and what would happen with different combinations of species and environmental changes.**
- What are the paths for transfer, and how can those paths be blocked or eliminated? **There are multiple paths for transfer, from bait buckets, to floods, to aquarium releases, and others. The likelihood of biota transfer via interbasin transfer of water can be reduced to next to nothing using chemical treatment, while the likelihood of biota transfer via other means approaches certainty. Reducing the likelihood via these other means is far more problematic.**

Glossary

amoebae. Any of a large genus (*Amoeba*) of naked rhizopod protozoans with lobed and never anastomosing pseudopodia and without permanent organelles or supporting structures that are widely distributed in fresh and salt water and moist terrestrial environments.

anadromous. A term describing the life history strategy of organisms that spend most of their life at sea and migrate to freshwater to spawn.

anthropocentric. Interpreting or regarding the world in terms of human values and experiences.

anthropogenic. Human impact on nature.

aquaculture. The culture of aquatic species under semi-controlled or controlled conditions.

bait bucked transfer. The transport and subsequent release of aquatic biota, through activities associated with sportfishing, into a watershed where it previously was absent.

ballast. A heavy substance (water) used to improve stability in a ship.

benthic invertebrates. Animals having no internal skeleton or backbone, found on the bottom of rivers and oceans.

biota. Living organisms.

Bowbells Block. An area of 57,000 acrcs near Crosby and Mohall in the northern part of North Dakota once considered a possible large-scale irrigation site. However, topographical, geological studies, and irrigation experiments from 1948-1950 revealed the area to be unsuitable for irrigation. The disqualification of the area was one of the first major setbacks to Garrison Diversion.

cichlids. Perch-like fish in the family Cichlidae, native to Africa, Asia, South America, Central America and southern North America.

ciliates. Any of a subphylum (Ciliophora) of ciliate protozoans.

clariids. Asian and African walking catfish, family Clariidae.

cobitids. Any fish, commonly referred to as loaches, in the family Cobitidae, native to Asia and Europe.

colpoda. A genus of small flattened reniform holotrichous freshwater ciliates.

conspecifics. Of the same species.

crustaceans. Any of a large class (Crustacia) of mostly aquatic arthropods that have a chitinous or calcareous and chitinous exoskeleton, a pair of often much modified appendages on each segment, and two pairs of antennae, such as the lobsters, shrimps, crabs, wood lice, water fleas, and barnacles.

cyprinids. Any minnow or carp-like fish in the family Cyprinidae, native to Asia, Africa, and North America.

diel. Involving a 24-hour period that usually includes a day and the adjoining night.

ecosystem. An assemblage of living organisms and non-living objects that form a stable interacting system. The interacting system of a biological community and its surroundings: a community of living organisms, found in a particular environment, interaction with their physical environment.

effluent. Outflowing.

endemic. Native to a particular location.

enteric redmouth. The disease caused by *Yersinia ruckeri* referring to the infection of the gastrointestinal tract and the inflammation accompanying the erosion of the jaw.

entrainment. Being pulled or brought along by water or other mediums.

exotic biota. An organism moved from a foreign country (i.e., on whose entire range is outside the country where found).

***ex post facto*.** Done, made, or formulated after the fact.

extirpate. Pull up by the root; to completely destroy.

extrinsic. External to.

fauna. Animals or animal life of a region, period, or geological stratum.

filamentous algae. An elongated thin series of cells attached one to another or a very long thin cylindical single celled aquatic nonvascular plants.

fish screen. The mechanical filtering device to remove fish or other debris from water.

flora. Plant life or plant life characteristics of a region, period, or special environment.

Garrison Diversion Conservancy District. Created in 1955, the GDCD was organized in North Dakota to work on the state level with the Bureau of Reclamation. It had the power to implement a mill levy, which added another source of revenue to help defray construction costs for diversion and irrigation. Only those areas in the state affected by diversion were included in the district.

genotype. The genetic constitution of a species.

gravid. Containing near mature eggs.

hybridization. To produce offspring through interbreeding two animals or plants of different races, breeds, varieties, species, or genera.

hydroacoustic. The movement of sound in water.

hydrology. Science dealing with the occurrence and distribution of water on and below the surface of the earth.

ichtyoplankton. Minute vertebrate animals having a fishlike form which float in great quantities near the surface of fresh or salt water.

indigenous species. Having originated and growing naturally in a particular region or environment.

interbasin transfer of water. The artificial withdrawal of water from one drainage basin and delivery for use in another drainage basin.

interspecific. Existing or arising between species.

intrinsic. Inherent or from within, generic.

introduced species. A plant or animal moved by humans from one place to another.

invasive species. Tending to spread and infringe on other species.

limnology, limnological. Scientific study of physical, chemical, meteorological, and biological conditions in fresh waters.

loricariids. Armored catfish of the family Loricariidae, native to South America.

low gradient streams. Water courses with gentle slopes.

macrocysts. Large cysts (capsulelike membrane).

macrophyte beds. Macroscopic plant life of a body of water.

metres, kilometres. Units of length in metric system of units. One metre equals 3.281 feet. One kilometre equals 1000 metres.

morphology. A branch of biology related to the form and structure of organisms. Also, topographic features produced by erosion.

motile organism. An organism capable of/and exhibiting movement.

necrosis. Death of a cell or group of cells (e.g. tissue) and accompanying degenerative changes as a result of disease or injury.

niche. A habitat which supplies what is necessary for an organism or species to exist.

non-indigenous. The condition of a species being beyond its natural range or natural zone of potential dispersal; includes all domesticated and feral species and all hybrids except for naturally occurring crosses between indigenous species.

non-target organisms. Biota found in waters in common with species identified for eradication.

ornamental fish. Fish species used in the aquarium business.

paradigm. A system of structured or organized thought about an issue.

parasites. Organisms that live on, or in, a larger host organism (plant or animal) that derive some benefit from the host.

pathogen. An organism that causes disease (e.g., bacteria, viruses, protozoan and/or metazoan parasitic organisms).

pelagic zone. The open water habitat in lakes, rivers, and ponds.

physiographic zone. Zone based on physical geography.

Pick-Sloan Plan. In 1944, the two separate plans of the Corps of Engineers (Pick Plan) and the Bureau of Reclamation (Sloan Plan) were combined and incorporated into the Flood Control Act of 1944. Under the compromise, the Corps of Engineers was given responsibility for determining the location of Missouri River main stream reservoirs and their capacity for flood control and river navigation. The Bureau of Reclamation was responsible for determining the location and capacities of minor reservoirs, the probable extent of irrigation of the Missouri River basin states, and the downstream depletion due to irrigation. The plan also divided the Missouri basin into subdivisions that allotted five reservoirs to the Corps and three to the Bureau, including the Garrison reservoir.

piscivorous. A descriptive term for an organism that preys on fish.

poeciliids. Livebearing fish in the family Poeciliidae, native to South and Central America.

poikilothermic. An organism having variable body temperature that is usually slightly higher than the temperature of its environment. A cold-blooded organism.

***Polypodium hydriforme*.** A cnidarian parasite of the eggs of acipenseriform fishes (sturgeons and paddlefish).

pracademics. Individuals in academic positions that are also active in "practicing what they preach."

predator. Animal given to preying upon other specific species.

progenitor. A biological ancestral form.

propagules. Stuctures (cuttings, seed, spore) that allow an organism to continue to increase by sexual or asexual reproduction.

protozoans. Any of a phylum or subkingdom (Protozoa) of minute protoplasmic acellular or unicellular animals which have varied morphology and physiology and often complex life cycles which are represented in almost every kind of habitat.

quaternary glaciation. Most recent geological time period, beginning about two million years ago.

regurgitation. Throwing back.

riffles. Shallow fastwater stretches in streams.

risk analysis. A process to assess the level of risk involved with a certain action.

salt water intrusion. Intrusion of salt water from the ocean into coastal aquifers due to lowering of pressure in the aquifers.

spores. A primitive, usually unicellular, resistant or reproductive body produced by plants and some invertebrates and capable of development into a new individual, in some cases unlike the parent, either directly or after fusion with another spore.

substrates. The base on which an organism lives.

sump. A pit or reservoir serving as a drain or receptacle for water.

target organisms. Biota identified for special treatment such as eradication.

taxa. Classes of living organisms.

terrestrial riparian vegetation. Vegetation growing on river banks/land as opposed to in water.

theoretical constructs. Part of the body of science that has evolved beyond speculation, hypotheses, and postulates.

total dissolved solids. The total amount of salts and minerals dissolved in water.

transfer rate factor. A measure to express the magnitude of the water transfer system. It is the product of the route length and the volume of water transferred ($Km\text{-}Km^3.yr^{-1}$).

trematodes. A class (Trematoda) of parasitic flatworms (Phylum: Platyhelminthes) commonly referred to as flukes.

trophic. Of, relating to, or characterized by nutrition.

turbidity. Murkiness of water due to suspended sediments and other materials.

unicellular algae. Having or consisting of a single cell.

viscera. Collective term for the organs occupying the body cavity of animals.

watershed. The entire land and water surface area that contributes runoff water to a lake, river, groundwater supply, or coastal waterbody. A geographic area in which water, sediment, and other dissolved materials drain to a common outlet.

water interception. Water diversions.

young-of-the-year. Juvenile fish hatched during the current year.

Geographic Index

Abram Lake, southwest Ontario, Canada
Alberta province, west Canada
Amu Darya basin, Amu Darya River, southern Uzbekistan (FSU)
Appalachia, eastern USA
Arctic Ocean
Arizona state, southwest USA
Arkansas state, south central USA
Arrowwood Refuge, Stutsman county, east central North Dakota, USA
Ashkhabad basin, south Turkmenistan (Ashgabat) (FSU)
Assiniboine River, southern Manitoba, Canada; drains into Red River
Athabasca River, west Alberta province, Canada
Australia
Baby Lonetree reservoir (Mid-Dakota), North Dakota, USA
Beas River, Himachal Pradesh state, north India
Beas-Sutlej link, north India
Bemidji city, northern Minnesota, USA
Bering Refugium, northeast Asia and northwest Alaska, USA
Bering Strait, Alaska, USA; connects Bering Sea and Chukchi Sea
Big Stone Lake, west central Minnesota, USA
Bismarck city, Burleigh county, west central North Dakota, USA
Blindfold Creek, Lake of the Woods, Minnesota, USA/Canada border
Bois de Sioux River, west central Minnesota, USA
Bonneville basin, west central Utah, USA
Bottineau County, north central North Dakota, USA
Bowbells Block, northwestern North Dakota, USA
Bowstring Lake, east of Bemidji, northern Minnesota, USA
Brahmaputra River, south Asia (India, China, Bangledesh); drains into Ganges River
Brazos River, southwest USA; flows into Gulf of Mexico
Brazos River basin, Texas, USA
British Columbia province, southwest Canada
Brokenhead River, northeast Manitoba, Canada; flows into L. Winnipeg
Browns Valley county, west central Minnesota, USA
Buffalo River, west central Minnesota, USA
Buford city, North Dakota, USA
Burke County, northwest North Dakota, USA
Burntside Lake, Minnesota, USA
California state, west USA
California aqueduct, California, USA
Canada
Canadian River, southwest USA (New Mexico, Texas, Oklahoma)
Central Asia
Central flyways, central North America
Chicago Sanitary Canal, Chicago, Illinois, USA
China, People's Republic of
Chukuni River, Ontario, Canada
Churchill Falls Project, Newfoundland, Canada
Churchill-Nelson diversion, Manitoba, Canada
Churchill Project, Manitoba, Canada
Clearwater River, Minnesota, USA
Colorado state, southwest USA
Colorado River, southwest USA (Arizona, California, Nevada)
Colorado River basin, southwest USA
Crosby-Mohall Unit, northwest North Dakota, USA
Cutfoot Sioux Lake, Mississippi basin, USA
Des Lacs River, North Dakota, USA
Detroit Lakes city, Minnesota, USA
Devils Lake city, Ramsey county, North Dakota, USA
Devils Lake (Spirit Lake) lake, Ramsey and Benson counties, North Dakota, USA
Devils Lake basin, North Dakota, USA
Dickey County, southeast North Dakota, USA
Divide County, northwest North Dakota, USA
Dryden city, Ontario, Canada
Duluth-Superior Harbor, harbor on Lake Superior at cities of Duluth, Minnesota, and Superior, Wisconsin
Ear Falls dam, English River area, southwest Ontario, Canada
Eddy County, east central North Dakota, USA
Elm River, east central North Dakota, USA
English River, southwest Ontario, Canada
Eucumbene River, New South Wales, southwest Australia
Eva Lake, Ontario, Canada
Fisher Bay, west side of Lake Winnipeg, Canada
Florida state, southeast USA
Forest River, west central North Dakota, USA
former Soviet Union (FSU)
Fort Berthold Indian Reservation, west central North Dakota, USA

Fort Peck city, northeast Montana, USA
Fort Peck dam, northeast Montana, USA
Ganges River, south Asia (India and Bangledesh); flows into Bay of Bengal
Garrison Diversion, North Dakota, USA
Garrison city, McLean county, northwest North Dakota, USA
Garrison Reservoir (Lake Sakakawea), northwest North Dakota, USA
Glacial Lake Agassiz, north central USA and Canada
Goose River, east central North Dakota, USA
Grand Canal (Yun Ho) (Da Yunhe), east China; south from Beijing to Hangzhou
Grand Forks city, Grand Forks county, northeast North Dakota, USA
Great Basin, west Nevada, USA
Great-Dividing Range, mountains east coast Australia (Queensland, New South Wales, Victoria)
Great Falls, southeast Manitoba, Canada
Great Fergana Canal, central Asia
Great-Fish River (Groot-Vis), south South Africa; flows into Indian Ocean
Great Lakes, lakes Erie, Huron, Michigan, Ontario, and Superior; border USA and Canada
Great Plain, China
Great Plains, central North America
Grenora city, Williams county, northeast North Dakota, USA
Gujrath, northeast India
Gulf Coast, south central USA; coast along Gulf of Mexico
Gulf of Mexico basin, south central USA
Haihe River, China
Haryana state, northwest India
Harvey city, Wells County, central North Dakota, USA
Heart River project, north central North Dakota, USA
Himalaya Mountains, south Asia
Huaihe River, China
Hudson Bay, northeast Canada
Hwang-Ho River (Hwang-Hae, Huang He, Yellow), central China
Icelandic River, Manitoba, Canada; north of Winnipeg, flows into Lake Winnipeg
Illinois River, Illinois, USA
India
Indus River, south Asia (Pakistan and India); flows into Arabian Sea
Iowa state, central USA
Jackhead River, Manitoba, Canada; north of Fisher Bay, Lake Winnipeg
James Bay, Quebec, Canada
James Bay Diversion, Quebec, Canada
James River, eastern North Dakota, USA
James River Canal, Eddy County, North Dakota, USA
Jamestown unit, North Dakota, USA
Jammu-Kashmir state, north India
Kansas state, central USA
Kansas City city, border Kansas and Missouri, USA
Karakum Canal (Karakumskiy Kanal), southeast Turkemistan (FSU)
Karakum Desert (Karakumy; Kara Kum), central Turkemistan (FSU)
Kardarya basin (FSU)
Karnool-Cuddapah Canal, south India
Keewatin city, northern Minnesota
Kettle Falls Dam, northeast Manitoba, Canada; south of Hudson Bay
Lac Seul Reservoir, on the English River, southwest Ontario, Canada
Lake Ashtabula, central North Dakota, USA
Lake Audubon, west central Minnesota, USA
Lake Erie, Great Lakes, north central USA/Canada border
Lake Manitoba, Manitoba, Canada
Lake Frances Case
Lake Michigan, Great Lakes, north central USA/ Canada border
Lake Oahe, north central South Dakota, USA
Lake of the Woods, Minnesota, USA/Manitoba, Canada border.)
Lake Sakakawea (Garrison reservoir), northwest North Dakota, USA
Lake Superior, Great Lakes, north central USA/ Canada border
Lake Traverse, west central Minnesota/northeast South Dakota
Lake Winnipeg, Manitoba, Canada
LaMoure city, LaMoure County, southeast North Dakota, USA
Limestone Dam, south of Hudson Bay, northeast Manitoba, Canada; by Kettle Rapids
Lisbon city, Ransom County, southeast North Dakota, USA
Little Eagle Lake, southwest Ontario, Canada
Little Minnesota River
Little Mississippi River
Little Missouri River

Lonetree Dam, Wells County, central North Dakota, USA
Lonetree Reservoir, Wells and Sheridan counties, central North Dakota, USA
Long Lake, east Manitoba, Canada
Long Spruce Dam, south of Hudson Bay, northeast Manitoba, Canada; by Kettle Rapids
Longbow Creek, Lake of the Woods, Minnesota/ Canada border
Louisiana state, south central USA
Louisiana Territory, USA; all of South Dakota,
Nebraska, Kansas, Oklahoma, Iowa, Missouri, Arkansas; parts of Minnesota, Louisiana, Texas, New Mexico, Colorado, Wyoming, Montana
Mackenzie River, Northwest Territories, Canada; flows to Beaufort Sea, Artic Ocean
Maine state, northeast USA
Mandan Bluffs, between Bismarck and mouth of Little Missouri River, North Dakota, USA
Mannhaven city, west central North Dakota, USA
Manitoba province, south central Canada
Maple River, eastern North Dakota, USA
Marksvill city, Louisiana, USA
Maynard Lake, southwest Ontario, Canada
McClusky Canal, McLean and Sheridan counties, North Dakota, USA; from Lake Sakakawea to Lonetree Reservoir
McGregor River, Pacific drainage, British Colum bia, Canada
McGregor River diversion, west Canada
McHenry County, north central North Dakota, USA
McKenzie County, west central North Dakota, USA
Medicine Lake Reservoir, northeast Montana, USA
Michigan state, north central USA
Mid-Dakota Reservoir (Baby Lonetree), North Dakota, USA
Middle River, northwest Minnesota, USA
Midwest region, central USA
Minnesota, state, north central USA
Minnesota River, southern Minnesota, USA
Minnitaki Lake, southwest Ontario, Canada
Mississippi state, southern USA
Mississippi River, central USA; flows into Gulf of Mexico
Missouri state, central USA
Missouri Canal, North Dakota, USA
Missouri River, north central USA
Montana state, north central USA
Moscow City, Russia (FSU)
Mukutawa River, central Manitoba, Canada
Murray basin, southeast Australia
Murray River, southeast Australia; flows into Tasmin Sea, South Pacific Ocean
Murrumbidgee River, southeast Australia
Muscova basin (FSU)
Mustinka River, Minnesota, USA
Naryn basin, Kyrgyzstan and Uzbekistan (FSU)
Nebraska state, central USA
Nechako-Kemano diversion, British Columbia, Canada
Nelson River, central Canada; flows into Hudson Bay
Nesson project, North Dakota, USA
Newfoundland province, east Canada
New Rockford Canal, Wells and Eddy counties, North Dakota, USA; from Lonetree Reservoir to James River
New Rockford unit, Wells and Eddy counties, North Dakota, USA
New York Mills city, east central Minnesota, USA
North China Plain, China
North Dakota state, north central USA
Oak Creek, Manitoba, Canada
Oak Lake, southwest Ontario, Canada
Oakes unit, Dickey and Sargent counties, southeast North Dakota, USA
Ogoki Lake, south central Ontario, Canada
Ohio state, northeast USA
Ohio River, north central USA; flows into Mississippi River
Ontario province, south Canada
Orange-Fish tunnel, South Africa; connects Orange River and Great Fish River
Orange River (Oranjerivier), southern Africa; flows into Atlantic Ocean
Otter Tail River, central Minnesota, USA
Pacific Ocean
Pacific coast, western North America
Pakwash Lake, Chukuni River lake chain, Ontario, Canada
Parambikulam-Aliyas project, India
Park River city, Walsh County, northeast North Dakota, USA
Parsnip River, north Canada, Artic drainage
Pelican Lake, west central Manitoba, Canada
Pelican River, east central Minnesota, USA
Pembina River, northeast North Dakota, USA
Periyar diversion, India
Playgreen Lake, west central Manitoba, Canada
Punjab state, north India
Quebec province, east Canada
Rafferty-Alameda Dams, Souris River,

Saskatchewan, Canada
Rainy Lake, southwest Ontario, Canada
Rainy River, southwest Ontario, Canada and northern Minnesota, USA
Rajasthan Canal, northwest India
Rajasthan state, northwest India
Ramaganga diversion, India
Ravi River, northwest India
Red Lake, Chukuni River lake chain, Ontario, Canada
Red Lake, northern Minnesota, USA
Red Lake River, northwest Minnesota, USA
Red River basin, Minnesota/North Dakota border, north central USA
Red River of the North, Minnesota/North Dakota border, north central USA
Renville County, north central North Dakota, USA
Rhode Island state, east coast USA
Rivers City, southwest Manitoba, Canada
Roseau River, northeast Minnesota, USA
Rush River, North Dakota, USA
Russia (Russian Federation) (FSU)
Sacramento River., California, USA; flows into San Joaquin River
St. Joseph city, Missouri, USA
St. Lawrence River, northeast USA/southeast Canada border
St. Louis city, Missouri, USA
St. Norbert city, south of Winnipeg, Manitoba, Canada
San Joaquin River, central California, USA; flows into Sacramento River
Sand Lake, North Dakota, USA
Sandybeach Lake, southwest Ontario, Canada
Sarda Sahayak project, northern India
Sargent County, southeast North Dakota, USA
Saskatchewan River, east central Saskatchewan, Canada; flows into Lake Winnipegosis, Manitoba
Sespe Sanctuary, California, USA
Shellmouth Dam, southwest Manitoba, Canada
Sheyenne River, North Dakota, USA
Shoal Lake, southwest Manitoba, Canada
Sioux City, Iowa, USA
Sipiwesk Lake, northwest Manitoba, Canada
Sisquoc Sanctuary, California, USA
Sisseton city, South Dakota, USA
Snake Creek, northwestern North Dakota, USA
Snake Creek Reservoir, northwestern North Dakota, USA
Snake River, Minnesota, USA
Snowy Mountain, Victoria, southeast Australia
Snowy River, Victoria, southeast Australia
Souris Canal, north central North Dakota, USA
Souris-Red-Rainy River basin, central Canada/USA border area
Souris River, north central North Dakota, USA, and southwest Manitoba, Canada
South Africa
South Dakota state, north central USA
South Saskatchewan River, southern Alberta and Saskatchewan, Canada
Spirit Lake Sioux Reservation, Ramsey and Benson counties, northeast North Dakota, USA
Split Lake, northern Manitoba (northeast of Lake Winnipeg), Canada
Standing Rock Sioux Reservation, southwest North Dakota and northwest South Dakota, USA
Stevens Lake, northern Manitoba (northeast of Lake Winnipeg), Canada
Stump Lake, Nelson County, northeast North Dakota, USA
Stutsman County, east central North Dakota, USA
Sutlej River, India and Pakistan; flows into Indus River
Sykeston canal, Wells County, central North Dakota, USA
Tamarac River, northwest Minnesota, USA
Texas state, southern USA
Three Affiliated Tribes, west central North Dakota, USA
Thunder Bay city, north shore of Lake Superior, Ontario, Canada
Tongue River, North Dakota, USA
Toronto city, Ontario, Canada
Turtle River, northeast North Dakota, USA
Upper Missouri basin, north central USA
United States
Valley City, east central North Dakota, USA
Velva Canal, northwest from Lonetree Reservoir north through McHenry County to Bottineau County, North Dakota, USA
Volga basin, west Russia (FSU)
Volga River, west Russia (FSU); flows into Caspian Sea
Volga-Moscova Canal, west Russia (FSU)
Voyageur National Park, northeast Minnesota, USA
Ward county, northwest North Dakota, USA
Warren's Landing, outlet at north end of Lake Winnipeg, Canada
Wawanesa city, southwest Manitoba, Canada
Whitefish Bay, Lake of the Woods, Minnesota/Canada border

Wild Rice River, eastern North Dakota, USA
Williams County, northwest North Dakota, USA
Williston city, Williams County, northwest North Dakota, USA
Winnipeg city, Manitoba, Canada
Winnipeg River, southern Manitoba, Canada; flows into Lake Winnipeg
Wisconsin state, north central USA
Wyoming state, northwest USA
Yangtze River (Kiang) (Chang Jiang), central China
Yankton city, southeast South Dakota, USA
Yellowstone River, Montana, USA
Yun Ho (Grand Canal) (Da Yunhe), China

Species list

bairdiella *Bairdiella icistia*
beaver (Castoridae)
bee (Apoidea)
bighead carp *Hypophthalmichthys nobilis*
bitterling *Rhodeus sericeus*
black-footed ferret *Mustela nigripes*
blacknose dace *Rhinichthys atratulus*
brine shrimp *Artemia salina*
Brook stickleback *Culaea inconstans*
brown trout *Salmo trutta*
bullfrog *Rana catesbiana*
bullhead *Ictalarus* Sp.
calanoid copepods *Leptodora* Spp.
California condor *Gymnogyps californianus*
channel catfish *Ictalurus punctatus*
cicada (Cicadidae)
common carp *Cyprinus carpio*
cormorant *Phalacrocorax* Sp.
crappie *Pomoxis* Sp.
crayfish (Decapoda)
dragonflies (*Dythemis, Plathemis, Libellula, Tramea,* and *Gomoides*)
emerald shiner *Notropis atherinoides*
Eurasian watermilfoil *Myriophyllum spicatum*
fathead minnow *Pimephales promelas*
finescale dace *Phoxinus neogaeus*
gizzard shad *Dorosoma cepedianum*
golden orf *Leuciscus idus*
golden shiner *Notemigonus crysoleucas*
goldfish *Carassius auratus*
grass carp *Ctenopharyngodon idella*
green sunfish *Lepomis cyanellus*
gull *Larinae* Sp.
ide *Leuciscus idus*
Iowa darters *Etheostoma exile*
juvenile burbot *Lota lota*
lake sturgeon *Acipenser fulvescens*
longear sunfish *Lepomis megalotis*
longnose dace *Rhinichthys cataractae*
longnose gar *Lepisosteus osseus*
longnose sucker *Catostomus catostomus*
mallard ducks *Anas platyrhynchos*
mosquito *Culex* Sp.
mosquitofish *Gambusia affinis*
muskrat *Ondatra zibethica*
northern brook lamprey *Ichthyomyzon fossor*
northern pike *Esox lucius*
northern redbelly dace *P. eos*
orangemouth corvina *Cynoscion xanthulus*
paddlefish *Polyodon spathula*
pallid sturgeon *Scaphirhynchus albus*
pearl dace *Margariscus margarita*
pelican *Pelicanus* Sp.
piranhas *Serrasalmus* Sp.
raccoons *Procyon* Sp.
rainbow smelt *Osmerus mordax*
red shiners *Cyprinella lutrensis*
river carpsucker *Carpiodes carpio*
river ruffe *Gymnocephalus cernuus*
river shiner *Notropis blennius*
round goby *Neogobius melanostomus*
rudd *Scardinius erthrophthalmus*
salmon (Salmonidae) family
shortnose gar *Lepisosteus platostomus*
shovelnose sturgeon *Scaphirhynchus platyrhynchus*
silvery minnow *Hybognathus nuchalis*
smallmouth buffalo *Ictobus bubalus*
spiny water flea *Bythotrephes cederstroemi*
spottail shiner *Notropis spilopterus*
stonecat *Noturus flavus*
sturgeon species *Acipenser nudiventris*
tench *Tinca tinca*
tilapia *Tilapia* Sp.
trout perch *Percopsis omiscomaycus*
tubenose goby *Proterorhinus marmoratus*
Utah chub *Gila atraria*
walleye pike *Stizostedion vitreum*
wasps (*Polistes, Pepsis*)
weather loach *Misgurnus anguilicaudatus*
white bass *Morone chrysops*
white sucker *Catostomus commersoni*
yellow perch *Perca flavescens*
zander *Sizostedion lucioperca*
zebra mussel *Dreissena polymorpha*

Contributors

Choudhury, A. Department of Zoology, University of Toronto, Toronto, Ontario, Canada.

Dick, Terry A. Department of Zoology, University of Manitoba, Winnipeg, Manitoba, Canada.

Franzin, William G. Dr. Franzin is Director, Department of Fisheries and Oceans' Freshwater Institute, University of Manitoba, Winnipeg, Manitoba, Canada.

Givers, David R. David Givers completed a BS in biology from Moorhead State University, Moorhead, Minnesota, in 1973, and an MS in Natural Resources Management, Department of Agricultural Economics, North Dakota State University, Fargo, in 1989. Mr. Givers began his professional environmental career working for the Tri-College University, Center for Environmental Studies, Fargo, North Dakota. Currently, Mr. Givers holds several job titles, including Program Officer for the Director for the North Dakota Water Resources Research Institute, North Dakota State University, Fargo. Mr. Givers held a temporary appointment with the Department of Physics, Concordia College, Moorhead, Minnesota, where he taught Environmental Studies.

Hanke, Gavin F. Dr. Hanke, former graduate research assistant for Dr. Kenneth Stewart, Department of Zoology, University of Manitoba, Winnipeg, Manitoba, Canada, is now with the Department of Biological Sciences, University of Alberta, Edmonton, Alberta, Canada.

Kelly, Paul. Mr. Kelly is a former native of northwest Minnesota and North Dakota. He graduated from North Dakota State University with an MS in 1989. His thesis was entitled "Under the Ditch: Irrigation and the Garrison Diversion Controversy." Mr. Kelly became interested in North Dakota water issues while serving as an intern for the Agricultural Committee of the North Dakota legislature in 1987. He is currently completing a Ph.D. in American History at the University of Nebraska - Lincoln. Areas of interest include the history of agricultural and rural societies, and the history of science.

Koel, Todd M. Formerly a graduate student in the Department of Zoology, North Dakota State University, Fargo, Dr. Koel is employed by the Illinois Natural History Survey, Havana, Illinois.

Krenz, Gene. Mr. Krenz received his BA and MA from the University of Montana and did post-graduate studies at the University of Chicago. He is currently Program Director for the Red River Basin Board, Moorhead, Minnesota. Mr. Krenz is a former Director of Planning for the North Dakota State Water Commission and the Souris-Red-Rainy River Basins Commission. He was appointed statewide coordinator for the Interbasin Water Transfer Study Program by Governor George Sinner. Mr. Krenz is co-author of "A River Runs North: Managing an International River."

Leitch, Jay A. Dr. Leitch has been Director of the Interbasin Water Transfer Study Program since 1988. He was professor of Natural Resource/Agricultural Economics at North Dakota State University. He is now Dean of the College of Business, North Dakota State University, Fargo.

Ludwig, Herbert R. Mr. Ludwig completed a master's degree in Agricultural Economics at North Dakota State University under Dr. Jay A. Leitch. His thesis is entitled "Bait Bucket Transfer Potential Betwen the Mississippi and Husdson Bay Watersheds."

McCulloch, Bruce R. Mr. McCulloch is a former graduate research assistant to Dr. Kenneth Stewart, Department of Zoology, University of Manitoba, Winnipeg, Manitoba, Canada. He completed a thesis entitled "Dispersal of the Stonecat (*Noturus flavus*) in Manitoba and its Interactions with Resident Fish Species."

Padmanabhan, Dr. G. Dr. Padmanabhan is a Professor of Civil Engineering with the Department of Civil Engineering and Construction at the North Dakota State University, Fargo. He has taught several water resources engineering related courses over fifteen years and has directed about twenty-five master's theses and three Ph.D. theses in water resources engineering area. Dr.

Padmanabhan is an active member in several professional societies, such as the American Society of Civil Engineers, the American Water Resources Association, the International Water Resources Association and the American Geophysical Union. He is a member of the Technical Advisory Team, since its inception, of the Interbasin Biota Transfer Study Program at North Dakota State University, Fargo, North Dakota, USA.

Peterka, John J. Dr. Peterka is professor emeritus of Zoology, North Dakota State University, Fargo.

Richard, Don Dr. Richard is Dean, School of Engineering and Mines, University of North Dakota, Grand Forks, North Dakota, USA.

Rosvold, Karl. Mr. Rosvold, having received his master's degree in Civil Engineering, North Dakota State University, Fargo, North Dakota, is now employed with Pro Gold, Wahpeton, North Dakota.

Souter, B. Fish Health, Department of Fisheries and Oceans, Winnipeg, Manitoba, Canada.

Stewart, Kenneth W. Dr. Stewart is professor of Zoology, University of Manitoba, Winnipeg, Manitoba, Canada.

Tenamoc, Mariah J. Ms. Tenamoc has a BS and an MS in sociology from North Dakota State University, Fargo. Currently, she is a doctoral candidate in the Department of Sociology at South Dakota State University. She has been associated with the Interbasin Water Transfer Study Program since 1989, working as administrative assistant, graduate project assistant, and research associate. Her MS thesis, "The Sociology of Multidisciplinary Research: A Case Study," was conducted with the sponsorship of the IBWSP.

Zimmerman, Robert A. Dr. Zimmerman, parttime lecturer, Department of Civil Engineering, North Dakota State University, Fargo, North Dakota, is the Superintendent of Environmental Systems/Assistant Public Works Director, Moorhead Waste Water Treatment Facility, Moorhead, Minnesota.